ANTARCTICA

ANTARCTICA

INDIA'S JOURNEY TO THE FROZEN CONTINENT

S.Z. Qasim
Dinabandhu Sahoo

NIYOGI
BOOKS

Published by

NIYOGI BOOKS
D-78, Okhla Industrial Area, Phase-I
New Delhi-110 020, INDIA
Tel: 91-11-26816301, 49327000
Fax: 91-11-26810483, 26813830
email: niyogibooks@gmail.com
website: www.niyogibooksindia.com

Editor: Gita Rajan
Design: Nabanita Das / Niyogi Books

ISBN: 978-93-81523-74-2
Publication: 2013

Printed at: Niyogi Offset Pvt. Ltd., New Delhi, India

This book is dedicated to Late Mrs Indira Gandhi, the former Prime Minister of India who was the pioneering spirit behind India's Antarctic programme.

Antarctica:
India's Journey to the Frozen Continent

"What thrilling news that our Indian Expedition has landed in Antarctica. It is the fulfilment of one of my long standing wishes ... The entire area is of deep interest to us and of vital importance. This will take us forward to greater success in science."

- Indira Gandhi
The then Prime Minister of India

Contents

Pg 1: Moving towards the frozen continent.

Pg 2-3: Imperial Gaze: Is the mood inspection or introspection?

Pg 4-5: Penguins in Antarctica: Life on Ice? One Big party, all fun and frolic!

Pg 6-7: Breaking the Ice: Entering Antarctica cutting through the frozen Southern Ocean.

Below: Beware: I am, watching you! Do not destroy my home.

Preface

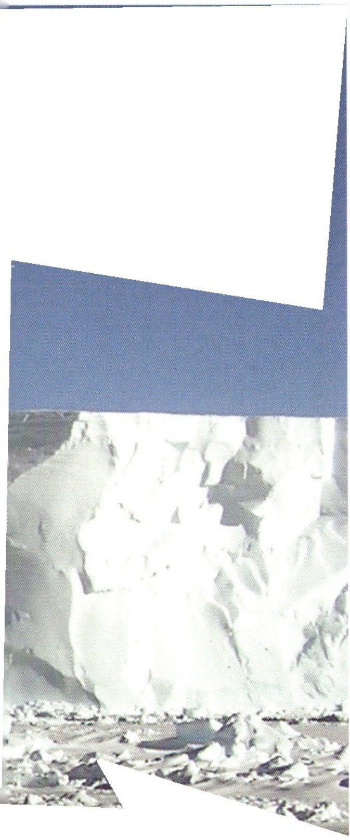

This book is about a journey, a mission, an initiative, a challenge, an adventure, a dream and finally a stepping stone for establishing the scientific capability of India in the world. This is an actual physical journey to a remote part of the world which we could not have even thought of earlier. This is not only about our journey, but also India's journey to Antarctica, which led to the foundation of India's polar research programme.

India launched its first ever scientific expedition to Antarctica in December 1981. Launching a scientific expedition to Antarctica was the foremost desire of Late Mrs. Indira Gandhi, the then Prime Minister of India who had a great interest for science. More than thirty years have passed and several expeditions have been successfully sent to Antarctica. Although, India has achieved several milestones, there is no first hand information available about the origin of the Antarctic programme during its formative years. This book not only describes the beauty of this enigmatic continent but also gives a first-hand experience of two different persons: one, a leader and the founder of India's Antarctic programme, and the other, a student who made his first ever overseas trip to the no-man's continent.

S.Z. Qasim
Dinabandhu Sahoo
9 January 2013

Fresh water for future!
Antarctica's Ice shelf.

Introduction

Fifty million years ago Antarctica had a temperate climate, evergreen forests and many more types of animals, but today this seventh continent is a white desert and the greatest wilderness in the world.

Antarctica is a continent for the future. It is a place for adventure, exploration, tourism, and a laboratory for science, unparalleled in terms of its vast mysteries. The unique climate, wind pattern, interaction of water in all its three forms namely ice, water and vapour, the short wave radiation from the sun, white-out conditions and the most unusual weather patterns offer great challenges to a scientist. What is so fascinating about Antarctica? Why should Antarctica be so cold and why is there so much ice accumulation in such a remote corner of the Earth? What are the consequences of this huge ice sheet in the surrounding oceans, atmosphere and its effect on global weather? What is stored under the ice and what do the large quantities of entrapped gases and other materials in the ice reveal about the past? Questions such as these may sound simple enough, but their answers are extremely complex, and to understand them, the scientific capability must be developed to a high standard. Besides, the quality of equipment and human endurance required for understanding the various phenomena over Antarctica, the creation of capabilities and infrastructure for conducting future experiments in a continent like Antarctica is in itself an immense scientific challenge.

Antarctica–the seventh continent of the Earth is known to be the coldest, driest, stormiest, windiest and the most inaccessible of all the other continents. The continent is commonly known as "White Desert" or "The

Penguins - Antarctica's native citizens.

Antarctica–the seventh continent of the Earth commonly known as "White Desert" is a land of mystery and wonder. The vast extent of ice, severe climatic conditions and the remoteness from the other parts of the globe, make it a virtual "Other World".

On the way to Antarctica: A view of the Ship through an iceberg.

Home of Blizzards" or "Penguins' Land" or "The Icy Continent" or "Southern Continent" the "Biggest Refrigerator in the world". This is not only a land of mystery and wonder but also the Earth's last great wilderness. The vast extent of ice, severe climatic conditions and the remoteness from the other parts of the globe, make it a virtual "Other World". A continent of extremes with nearly six months of continuous day and about six months of continuous night, the lowest temperature of -89.6˚C and wind speeds up to 190 km per hour during blizzards make this continent a unique place on Earth.

It is Earth's last frontier, the most inaccessible and least explored part of the world today. A continent of extremes and mysteries, Antarctica is completely different from any other part of the world. Going to Antarctica in those early years of expedition was like going to space. In other words, you can call it going to heaven or hell depending on your experience during the journey to the mysterious continent. It is the driest of all continents and also called "Cold Desert" where nearly 90% of the world's fresh water is stored in the form of ice. Although it is the fifth biggest continent, its largest permanent inhabitant is a wingless fly, but still Antarctica is known as Penguins' Land. Although it

Polar Circle the icebreaker ship of the First Indian Expedition.

has an area of fourteen million square kilometres (bigger than India and China put together) only 2000-3000 people live there during the summer season and only around 1000 during the extreme winter season. It is the least inaccessible as compared to all the other six continents. More than 98% of the continent is covered with ice. This is unlike the Arctic where people can obtain food; and there has been a native population of Eskimos living in the Arctic for thousands of years. The Antarctic, on the other hand, is totally devoid of an indigenous human habitation.

The average mean temperature in Antarctica is -35°C and the lowest recorded was -89.6°C. The wind gusts can exceed speeds of 300 km per hour. Add the wind chill factor; the temperature can plummet below -100°C. Although it has huge snow accumulations in its ice cap, the average snowfall is only 50 mm per year as the air is very dry. Today, Antarctica is a place of extremes.

India's journey began to this remote, isolated and extreme continent in search of knowledge and for unravelling its mystery. The First Indian Scientific Expedition to Antarctica was launched on 6th December 1981 led by Dr S.Z. Qasim an Internationally acclaimed Oceanographer. Since then every year an expedition has been launched by India until the present. In the year 1987, for the first time a student from an Indian University, Dinabandhu Sahoo, the co-author of this book, visited Antarctica. As you go through this book, you will come to know more about this continent from two different persons: one a leader and the other a student.

© Mike Usher

A common sight in Antarctica is a large mass of ice broken from the ice shelf floating around in the frozen ocean looking picturesque. The frozen southern ocean adds to the natural beauty of the landscape. Cold, yet uniquely beautiful and inviting.

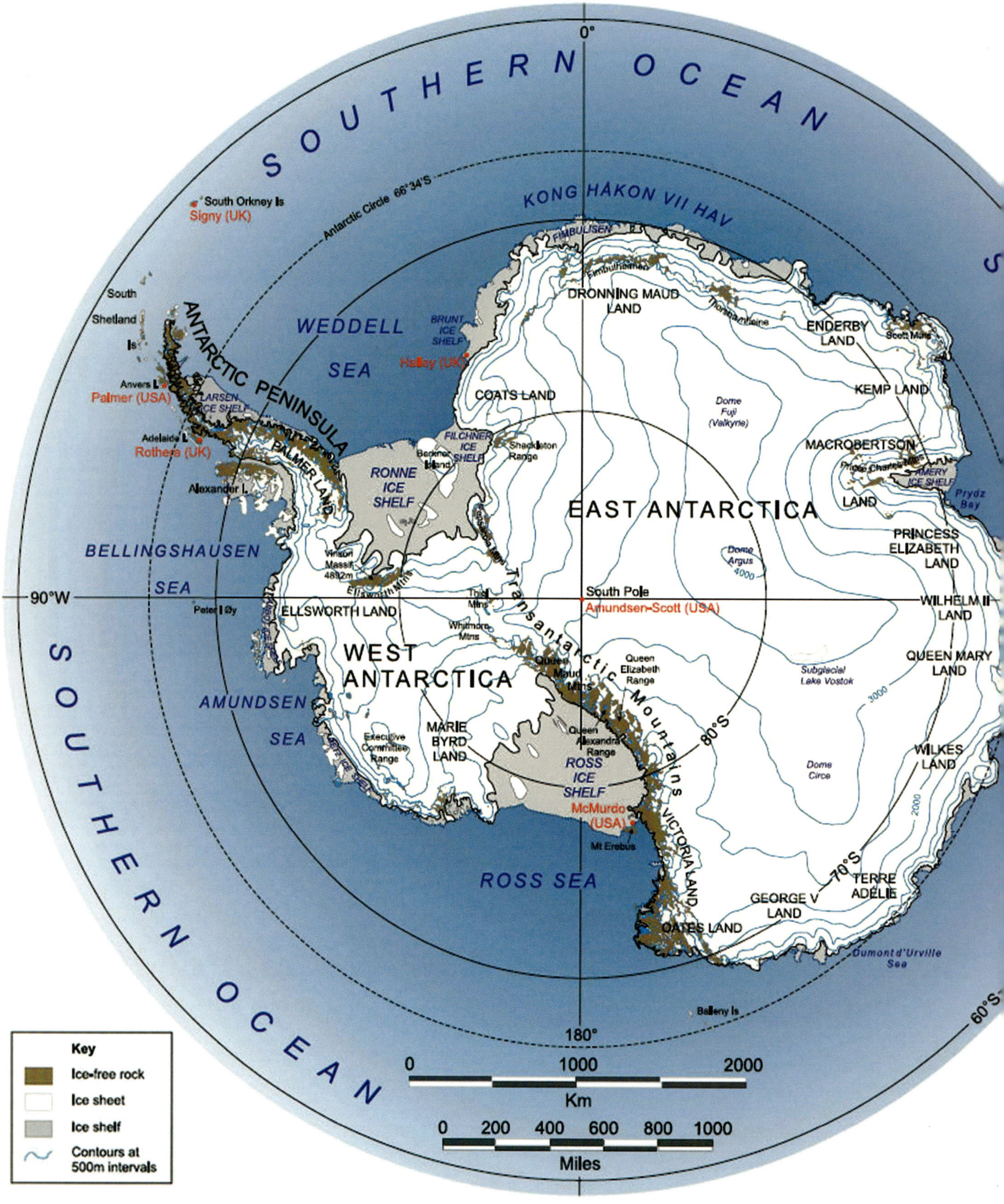

SOUTHERN OCEAN
0°
KONG HÅKON VII HAV
Antarctic Circle 66°34'S
South Orkney Is
Signy (UK)
South Shetland Is
ANTARCTIC PENINSULA
WEDDELL SEA
BRUNT ICE SHELF
Halley (UK)
FIMBULISEN
Fimbulheimen
DRONNING MAUD LAND
ENDERBY LAND
Scott Mts
KEMP LAND
Anvers I.
Palmer (USA)
LARSEN ICE SHELF
Adelaide I.
Rothera (UK)
PALMER LAND
Alexander I.
RONNE ICE SHELF
FILCHNER ICE SHELF
Berkner Island
Shackleton Range
COATS LAND
Dome Fuji (Valkyrie)
MACROBERTSON LAND
Prince Charles Mts
AMERY ICE SHELF
Prydz Bay
EAST ANTARCTICA
BELLINGSHAUSEN SEA
Vinson Massif 4892m
Ellsworth Mts
Dome Argus
4000
PRINCESS ELIZABETH LAND
90°W
Peter I Øy
ELLSWORTH LAND
Thiel Mts
South Pole
Amundsen-Scott (USA)
WILHELM II LAND
Whitmore Mts
Transantarctic Mountains
WEST ANTARCTICA
Queen Maud Mts
Queen Elizabeth Range
QUEEN MARY LAND
Subglacial Lake Vostok
3000
AMUNDSEN SEA
MARIE BYRD LAND
Executive Committee Range
Queen Alexandra Range
80°S
ROSS ICE SHELF
WILKES LAND
Dome Circe
2000
McMurdo (USA)
Mt Erebus
VICTORIA LAND
ROSS SEA
70°S
TERRE ADÉLIE
GEORGE V LAND
OATES LAND
Dumont d'Urville Sea
Balleny Is
180°
60°S
SOUTHERN OCEAN
Key
Ice-free rock
Ice sheet
Ice shelf
Contours at 500m intervals
0 1000 2000
Km
0 200 400 600 800 1000
Miles

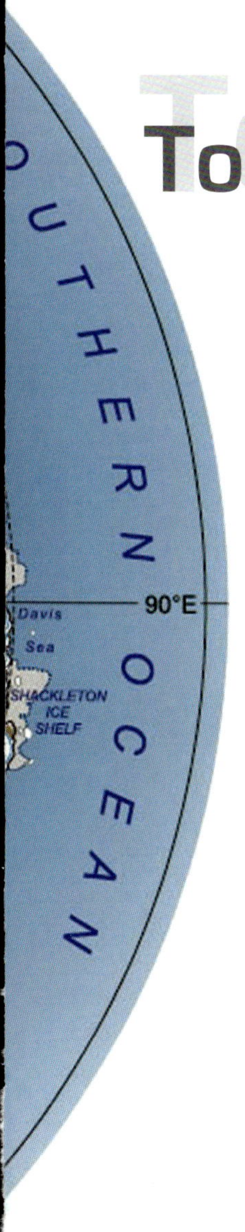

Topography

Antarctica is an island continent situated more or less centred on the South Pole. The coastline is about 14,000 miles in perimeter. It is a pulsating continent. During the winter, with its ice cap, Antarctica is roughly circular with a diameter of about 4500 kilometres (2800 miles) and an area of fourteen million square kilometres (5.4 million square miles). If all the ice was removed from Antarctica, the landmass below the ice would be reduced to half, measuring only seven million square kilometres (2.7 million square miles). It is larger than the United States of America and Europe combined but less than half the size of Africa and twice the size of Australia. Surrounding the continent is a frozen ocean known as the Southern Ocean or Antarctic Ocean. During the Antarctic winter the Southern Ocean is frozen into ice covering an area of nearly nineteen million square kilometres whereas by the end of peak summer the sea-ice is reduced to an area of nearly 2.65 million square kilometres. Antarctica is divided both geographically and geologically into two natural provinces.

1. Lesser Antarctica or Western Antarctica, known geologically as Andean Province.
2. Greater Antarctica or Eastern Antarctica, geologically called Gondwana Province.

Facing page:
Transantarctic Mountains,
West Antarctica,
East Antarctica

ANTARCTICA
A NEW AGE OF EXPLORATION

NATIONAL GEOGRAPHIC

Extremes of climate and terrain found on no other continent confront all who venture to Antarctica. Almost a century ago iron men named Amundsen, Shackleton, and Scott raced exhaustion, starvation, frostbite, and each other to the South Pole. Planting Norway's flag, Amundsen won. Today's goal is knowledge: Satellite and ground-based observations help scientists understand the continent and its global climate impacts. The scope and detail achieved by Radarsat's recent Antarctic mapping missions "was once unimaginable," according to Ghassem Asrar, NASA's associate administrator for earth science. "Scientists and engineers have literally created new ways to see the remotest reaches of the planet."

Radarsat Fills in the Blanks

Maps of Antarctica's interior remained mostly white blanks into the mid-1980s. Satellites using visible light had produced detailed surface images, but their angles of view excluded more than 1.2 million square miles poleward of about 82° south latitude. Then in 1997 the Canadian Space Agency (CSA) rotated its Radarsat I satellite in orbit, directing its radar sensor toward the South Pole. The first Antarctic Mapping Mission, a joint project of CSA and NASA in the U.S., imaged the entire continent in just 18 days at a resolution many times finer than any other single satellite survey. Compiling Radarsat's 4,500 scans into the digital mosaic seen here required two years. Dark areas on the map indicate smooth ice and fine-grained surfaces, like new snow, that don't scatter the radar beam. Coarse surfaces—old granular snow, rough ice, and crevasses—break up the beam and appear bright. During its many passes over the continent the satellite scanned the surface in strips, producing the faint starburst effect that patterns the image. In 2000, Radarsat remapped Antarctica's coast and most of its fast-moving glaciers, astonishing researchers with detailed images of recent change.

RADARSAT I MOSAIC CREATED BY KENNETH JEZEK, BYRD POLAR RESEARCH CENTER, OHIO STATE UNIVERSITY, BASED ON 1997 DATA FROM CANADIAN SPACE AGENCY, NASA, NASA/JPL, VEXCEL CORPORATION, AND ALASKA SAR FACILITY

ANTARCTIC PENINSULA
A mountain range welded to clusters of islands by a relatively thin coat of ice, this 800-mile-long peninsula is popular with penguins and other seabirds, including gulls, skuas, and petrels and provides important habitat for several species of seals.

A DIFFERENT WORLD
Able to survive months of dry, frigid darkness, lichens and mosses are Antarctica's dominant plants. The largest land animal is a wingless midge about a half inch long.

Surface Elevation

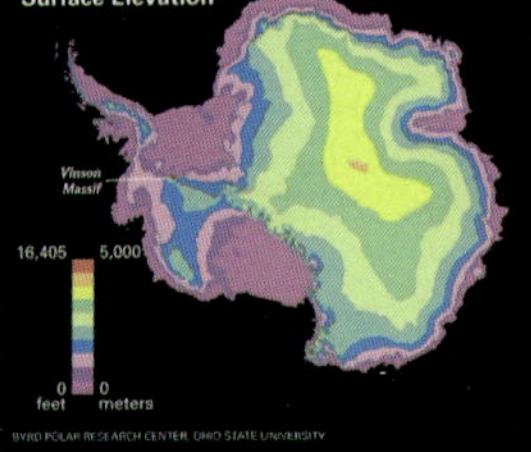

ELEVATION OF THE ICE SHEET

Many mountaintops rise higher than Antarctica's highest point—Vinson Massif, 16,067 feet—but with an average elevation of 8,000 feet, the continent ranks as Earth's highest. Asia, its closest competitor, averages 3,000 feet. Roughly dome shaped, the ice sheet conceals much of the bedrock relief below. The 1,800-mile-long Transantarctic Mountains rival the Rockies in height, but only the peaks break through the ice.

Ice Sheet Thickness

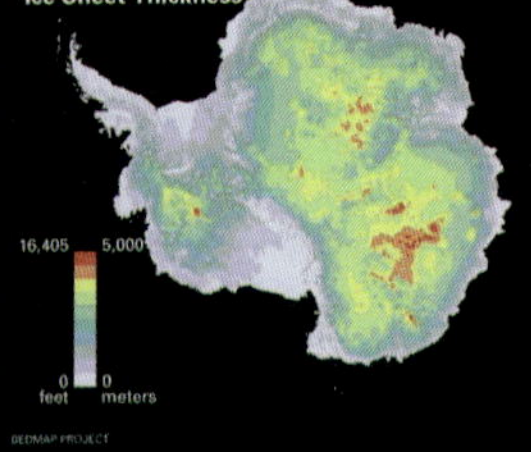

MEASUREMENTS OF A PARADOX

Ninety percent of the world's ice and 70 percent of the world's fresh water are found here, yet most of Antarctica is truly a desert. The snow equivalent of less than three inches of rain—a precipitation rate close to that of the Sahara—falls over the high interior of the continent each year. But snow and ice have been slowly accumulating on Antarctica for millions of years. More than 15,600 feet deep at its thickest, the mean depth of the ice exceeds 6,600 feet.

Ice Flow Velocity

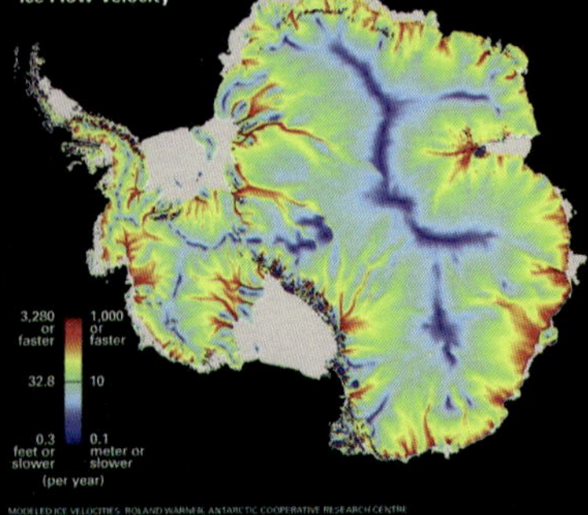

◄ICE ON THE MOVE

Glaciologists once thought that ice motion in Antarctica's interior was slow and relatively uniform, with just a few fast-moving outlet glaciers and ice streams drawing ice from the interior down to the sea. A computer model based on new satellite data suggests a more intricate ice-movement pattern (left). Like rivers, coastal ice flows appear to be fed by complex systems of tributaries that penetrate hundreds of miles into major drainage basins.

ULTIMATE WINDS►

Katabatic winds—cold air pouring down glacial slopes—often blow at 80 miles an hour and can exceed 180 miles an hour. Offshore, circumpolar winds and currents push against sea ice that rings Antarctica each winter, driving the frozen ocean counterclockwise around the continent.

Sea Ice Movement and Wind Flow

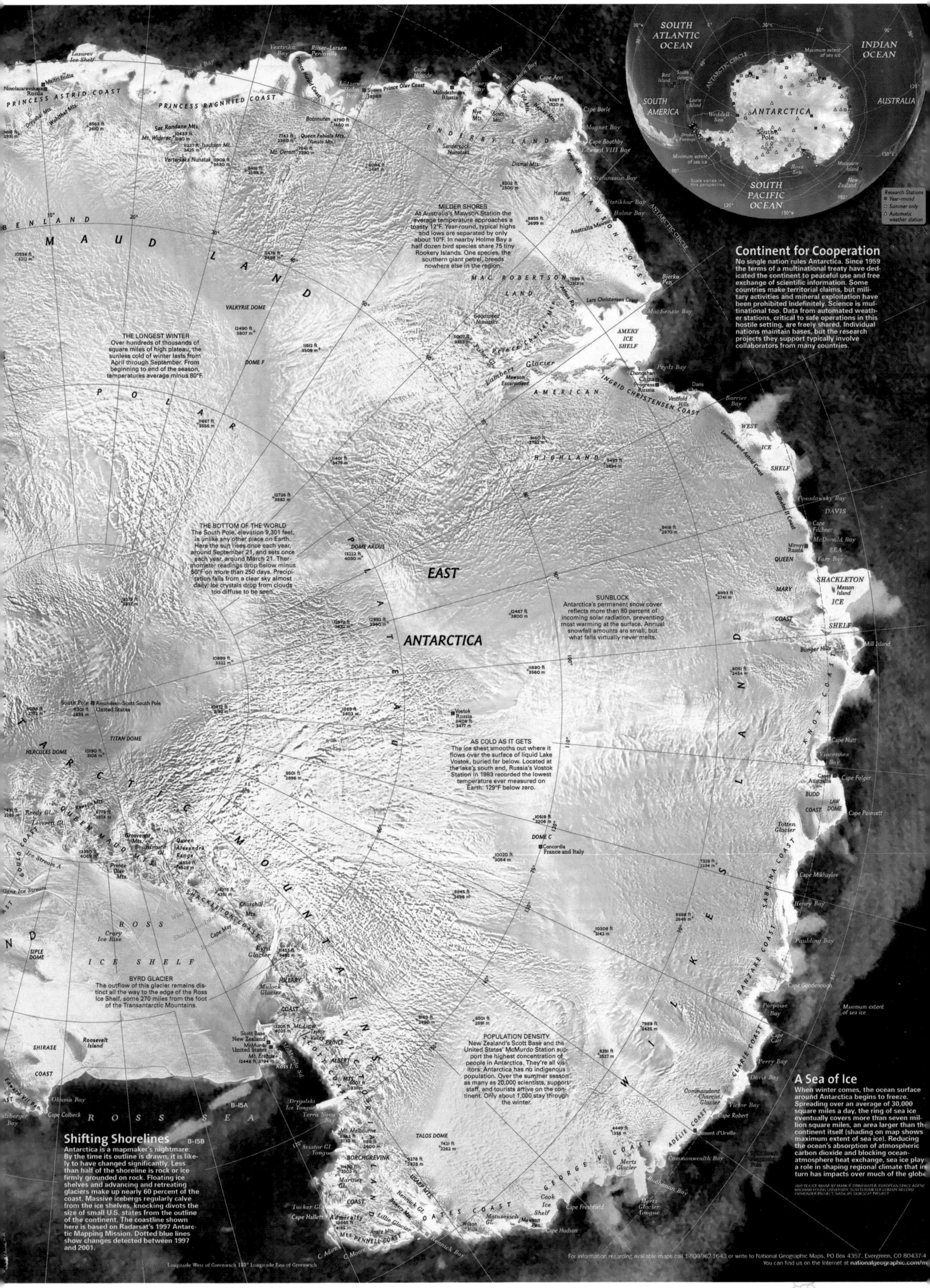
SOUTH ATLANTIC OCEAN
INDIAN OCEAN
SOUTH AMERICA
ANTARCTICA
South Pole
AUSTRALIA
SOUTH PACIFIC OCEAN
Weddell Sea
Ross Sea
ANTARCTIC CIRCLE
Maximum extent of sea ice
Minimum extent of sea ice
Research Stations
Year-round
Summer only
Automatic weather station
Continent for Cooperation
No single nation rules Antarctica. Since 1959 the terms of a multinational treaty have dedicated the continent to peaceful use and free exchange of scientific information. Some countries make territorial claims, but military activities and mineral exploitation have been prohibited indefinitely. Science is multinational too. Data from automated weather stations, critical to safe operations in this hostile setting, are freely shared. Individual nations maintain bases, but the research projects they support typically involve collaborators from many countries.
MILDER SHORES
At Australia's Mawson Station the average temperature approaches a toasty 12°F. Year-round, typical highs and lows are separated by only about 10°F. In nearby Holme Bay a half dozen bird species share 75 tiny Rookery Islands. One species, the southern giant petrel, breeds nowhere else in the region.
THE LONGEST WINTER
Over hundreds of thousands of square miles of high plateau, the sunless cold of winter lasts from April through September. From beginning to end of the season, temperatures average minus 80°F.
THE BOTTOM OF THE WORLD
The South Pole, elevation 9,301 feet, is unlike any other place on Earth. Here the sun rises once each year, around September 21, and sets once each year, around March 21. Thermometer readings drop below minus 50°F on more than 250 days. Precipitation falls from a clear sky almost daily: Ice crystals drop from clouds too diffuse to be seen.
SUNBLOCK
Antarctica's permanent snow cover reflects more than 80 percent of incoming solar radiation, preventing most warming at the surface. Annual snowfall amounts are small, but what falls virtually never melts.
AS COLD AS IT GETS
The ice sheet smooths out where it flows over the surface of liquid Lake Vostok, buried far below. Located at the lake's south end, Russia's Vostok Station in 1983 recorded the lowest temperature ever measured on Earth: 129°F below zero.
BYRD GLACIER
The outflow of this glacier remains distinct all the way to the edge of the Ross Ice Shelf, some 270 miles from the foot of the Transantarctic Mountains.
POPULATION DENSITY
New Zealand's Scott Base and the United States' McMurdo Station support the highest concentration of people in Antarctica. They're all visitors: Antarctica has no indigenous population. Over the summer season as many as 20,000 scientists, support staff, and tourists arrive on the continent. Only about 1,000 stay through the winter.
EAST ANTARCTICA
MAUD LAND
POLAR PLATEAU
TRANSANTARCTIC MOUNTAINS
WILKES LAND
ENDERBY LAND
MAC ROBERTSON LAND
AMERICAN HIGHLAND
PRINCESS ASTRID COAST
PRINCESS RAGNHILD COAST
PRINCE OLAV COAST
MAWSON COAST
INGRID CHRISTENSEN COAST
QUEEN MARY COAST
KNOX COAST
BUDD COAST
SABRINA COAST
BANZARE COAST
CLARIE COAST
ADÉLIE COAST
GEORGE V COAST
OATES COAST
PENNELL COAST
BORCHGREVINK COAST
SCOTT COAST
SHACKLETON COAST
AMERY ICE SHELF
WEST ICE SHELF
SHACKLETON ICE SHELF
ROSS ICE SHELF
ROSS SEA
DAVIS SEA
Lambert Glacier
Prince Charles Mts.
DOME ARGUS
DOME F
DOME C
VALKYRIE DOME
TITAN DOME
HERCULES DOME
TALOS DOME
SIPLE DOME
South Pole
Amundsen-Scott South Pole United States
Vostok Russia
Concordia France and Italy
Scott Base New Zealand
McMurdo United States
Mt. Erebus
Mirnyy Russia
Mawson Australia
Casey Australia
Dumont d'Urville
Syowa Japan
Novolazarevskaya Russia
Maitri India
Zhongshan China
Progress Russia
Davis Australia
A Sea of Ice
When winter comes, the ocean surface around Antarctica begins to freeze. Spreading over an average of 30,000 square miles a day, the ring of sea ice eventually covers more than seven million square miles, an area larger than the continent itself (shading on map shows maximum extent of sea ice). Reducing the ocean's absorption of atmospheric carbon dioxide and blocking ocean-atmosphere heat exchange, sea ice plays a role in shaping regional climate that in turn has impacts over much of the globe.
Shifting Shorelines
Antarctica is a mapmaker's nightmare: By the time its outline is drawn, it is likely to have changed significantly. Less than half of the shoreline is rock or ice firmly grounded on rock. Floating ice shelves and advancing and retreating glaciers make up nearly 60 percent of the coast. Massive icebergs regularly calve from the ice shelves, knocking divots the size of small U.S. states from the outline of the continent. The coastline shown here is based on Radarsat's 1997 Antarctic Mapping Mission. Dotted blue lines show changes detected between 1997 and 2001.
B-15A
B-15B
For information regarding available maps call 1-800-962-1643 or write to National Geographic Maps, PO Box 4357, Evergreen, CO 80437-4
You can find us on the Internet at nationalgeographic.com/m
Longitude West of Greenwich 180° Longitude East of Greenwich

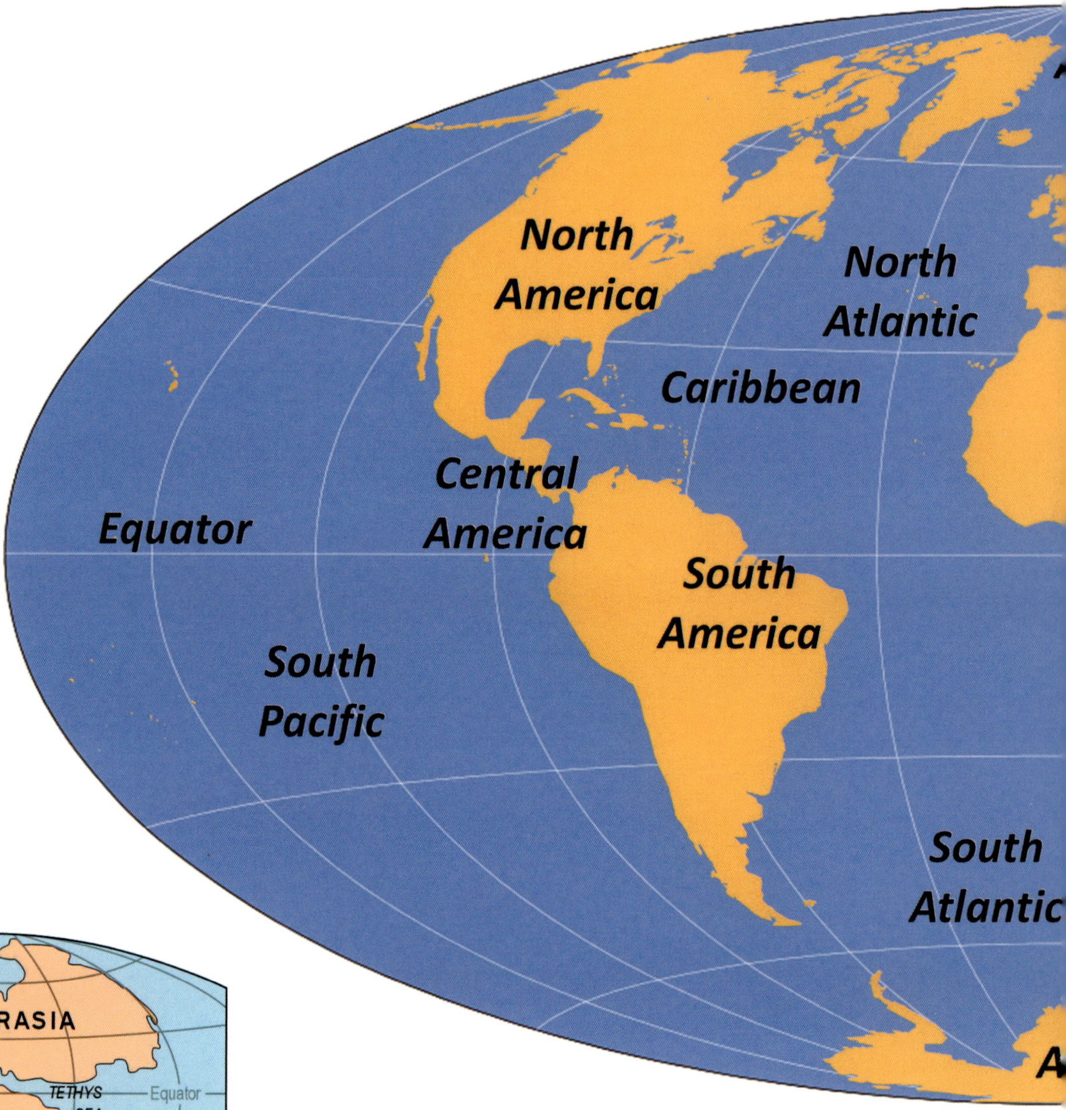

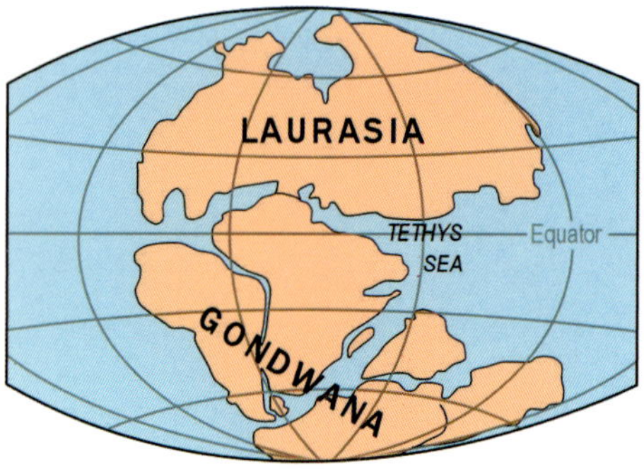

TRIASSIC
200 million years ago

Millions of years ago, Antarctica, India, Africa, Australia, New Zealand and South America were joined together and it was called the supercontinent "Gondwanaland". Some 180 million years ago, it was pulled apart by forces within the Earth's crust and started drifting. The continental drift continues today and the southern hemisphere sea floor spreads 1-6 centimetres every year.

Eastern Antarctica is twice the size of Western Antarctica and geologically much older.

Antarctica is one of the fragments of former super continent Gondwanaland, which began to break into several parts 160 million years ago. In other words, once upon a time India, Australia, New Zealand, South America and Africa were joined together as a common landmass. The break up with Australia began about eighty-six million years ago. Antarctica moved to its present position thirty million years ago. Presently Antarctica is separated from the other continents by the stormy and turbulent waters of the Southern Ocean. The nearest landmass to Antarctica is Cape Horn at the very tip of South America and this is still 965 kilometres away. The distances from Antarctica to Australia, Africa and India are 2500 km, 4000 km and 12000 km respectively.

Above: After millions of years this is how the world looks like.

Facing page: Once upon a time all the continents were joined together.

Pages 22-23: Geographical boundaries of Antarctica.
Courtesy: National Geographic.

Snow, Blizzards, Ice caps and Icebergs

Snow, Blizzards, Ice caps and Icebergs

Antarctica holds about 90 percent of the world's ice which constitutes around 70% of all Earth's fresh water. This represents an enormous volume of water about 24.5 million cubic kilometres (5.9 million cubic miles) as 98% of the continent is covered with ice leaving only 2% of land area which can be seen only during the Antarctic summer. Since it is the driest continent in the world, there is always very low humidity in the air.

Snow Blizzards

The most difficult weather conditions in Antarctica are *Katabatic* winds and *Blizzards*. (Katabatic winds are most commonly found blowing out from the large and elevated ice sheets of Antarctica and Greenland. The build-up of high density cold air over the ice sheets and the elevation of the ice sheets brings into play enormous gravitational energy. Where these winds are concentrated into restricted areas in the coastal valleys, the winds blow well over hurricane force. Blizzards are a combination of strong wind and falling or blowing

Both above: Post blizzard effect, frozen but still happy.

Facing page: A typical blizzard condition in Antarctica with low visibility (at *Dakshin Gangotri*). Sahoo standing in centre.

If by mistake you are caught in a blizzard it becomes the most harrowing experience in life. The snow particles in the air hit your body like pins or razor blades. A bearded face accumulates huge amount of snow which subsequently turns into ice which is very difficult and painful to remove from the beard. It is always advisable to stay inside the station during a blizzard.

A typical blizzard condition in Antarctica showing extremely low visibility.

Facing page: Impact of frostbite difficult to cure (left); the beard after blizzard (right).

Right: A husky after blizzard.

Below: A husky before blizzard.

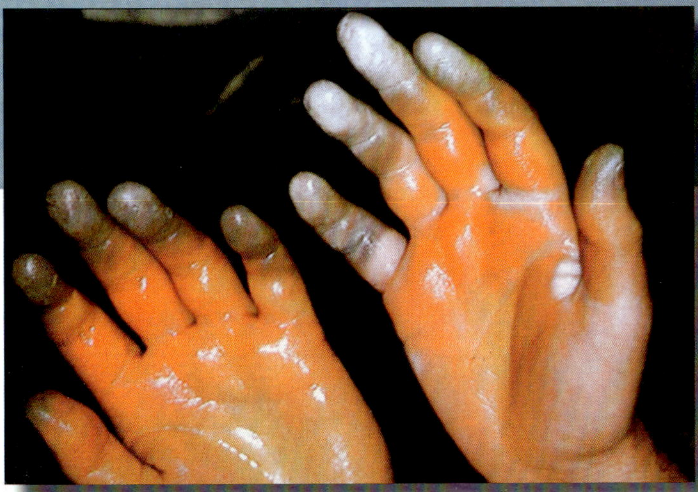

snow. During blizzards the visibility can become less than a metre. The wind speed can go up to 190 km/h making life very difficult. It can physically lift and throw a human several metres away. A blizzard can last for several days making life difficult. Tents are blown away, communication system gets disconnected, some equipments stop working and the most difficult thing is confinement to very limited spaces inside the cabins.

Facing page: Glacier in the background at Schirmacher Oasis (above); looking up. How to climb this icy mountain! (below).

Below: Measuring his height standing before the glacier at *Maitri.*

Lambert glacier which is around 40 km wide and flows for over 400 kilometres is not only the largest in Antarctica but the largest in the world as well. If all the ice in Antarctica will melt then the sea level in the world would rise between 50-60 metres (160-200 feet).

The snow is carried by wind from place to place and gets deposited in layers. The ice that covers the continent is called the ice sheet or continental ice or polar ice which is formed by the deposit of layers of snow over a period of thousands of years. Antarctica is the highest of all the continents. Its average elevation is 2300 metres (7500 ft). The minimum thickness of ice is 800 metres whereas the maximum depth of ice is 4,800 metres (15,700 feet) which is the thickest ice in the world. The average thickness of the ice-cap that covers the continent is 2400 metres. The ice-cap is thick in East Antarctica and relatively thin in the West Antarctic region. The ice flows downhill from the high points of the ice-cap towards the coast under the influence of its own weight. Antarctica is drained by vast systems of glaciers and ice streams. (A glacier is a large persistent body of ice that forms where the accumulation of snow exceeds its ablation (melting and sublimation) over many years, often centuries to many thousands of years. A glacier is at least 0.1 km^2 in area and 50 m thick.

Icebergs

There are tens of thousands of majestic icebergs which float in the Antarctic Ocean. (An Iceberg is a large piece of freshwater turned ice that has broken off from an ice shelf or glacier and floats in the sea). These floating masses of ice are the fragments of about 600,000 million tons of ice which drain from the continental ice-sheet each year. Icebergs occur in a variety of shapes and sizes. They are shaped by the sea and wind into myriads of graceful curves and these turn into towering pinnacles. Some are gleaming white, while others are bottle glass green or pink or golden as they reflect the setting sun. An average iceberg has a life of perhaps four years, during which time it will migrate to warmer seas. Eventually these large bergs break up and turn into

water. The largest iceberg on record was an Antarctic tabular iceberg of over 31,000 square kilometres or 335 by 97 kilometres (12,000 sq miles or 208 by 60 miles) sighted 240 km west of Scott Island, in the South Pacific Ocean, by the *USS Glacier* on November 12, 1956. This iceberg was larger than the size of Belgium. Antarctic icebergs, unlike those from the Arctic, are not usually carved from glaciers, but consist of portions that have broken away from the great ice shelves which surround much of the Antarctic Continent. When an iceberg melts, it makes a fizzing sound jokingly called "Bergie Seltzer". This sound is made when compressed air bubbles trapped in the iceberg pop up. The bubbles come from the air trapped in snow layers that later become glacial ice.

The majestic icebergs close-up!

Antarctic icebergs, unlike those from the Arctic, are not usually carved from glaciers, but consist of portions that have broken away from the great ice shelves which surround much of the Antarctic Continent.

© Mike Usher

There are tens of thousands of majestic icebergs which float in the Antarctic Ocean. Icebergs occur in a variety of shapes and sizes. They are shaped by the sea and wind into a myriad of graceful curves and these turn into towering pinnacles. Some are gleaming white, while others are bottle green or pink or golden as they reflect the setting sun.

Tip of the iceberg

The density of pure ice is about 920 kg/m^3, and that of sea water about 1025 kg/m^3. Typically only one-ninth of the volume of an iceberg is above water. The shape of the underwater portion can be difficult to judge by looking at the portion above the surface. This has led to the expression "tip of the iceberg", for a problem or difficulty that is only a small or miniscule manifestation of a larger problem.

The sea ice is perhaps the most enthralling of all. Its beauty changes with the seasons. It forms in the autumn, expands widely in winter and begins to break up in spring. In summer it drifts as pack-ice until much of it melts away as it moves into somewhat warmer waters to the north. One of the most wonderful sights is the formation of the next generation of ice. Millions of small *pancake-like* floes about a metre in width stretch to the horizon.

Facing page: Ice formation at the onset of winter. Greasy ice.

Sea water is freezing to ice on the onset of winter.

Facing page: Pancake Ice.

Breaking of ice at the early setting of summer

Antarctica is home to more than 150 lakes that lie at the base of the continental ice sheet. Lake Vostok is the largest of these sub glacial lakes. It was discovered beneath Russia's Vostok Station in 1996. Priyadarshini Lake (named after late Mrs Indira Priyadarshini Gandhi, a former prime minister of India) is located at Schirmacher Oasis facing India's permanent station *Maitri*.

Plants in Antarctica

Unlike the Arctic regions, species diversity in Antarctica is low. The scarcity of plant and animal life in Antarctica is due to extreme climatic conditions, lack of suitable habitats and probably also its isolation being a barrier to dispersal. The snow lies on the ground almost the entire year. Only 2% of landmass gets exposed during the summer and this is the reason for the scanty vegetation in Antarctica. The factors that produce this low abundance of plant life in Antarctica are aridity, strong winds, absence of sufficient sunshine and frozen conditions for nearly six months with continuous daylight and almost the same period of continuous darkness. Moreover, the soil is of poor fertility being deficient in nutrients. Thus, one does not find trees, grass or common large animals in the continent.

Only two native flowering plants of very small size, a grass named *Deschampsia antarctica* and a pearlwort *Colobanthus quitensis* (known as Antarctic Pink) have been reported from the western side of the Antarctic Peninsula.

The Sub-Antarctic islands with a little more favourable climatic conditions and better soil fertility support nearly thirty species of flowering plants and ferns. In contrast, the Arctic region supports nearly one hundred species of flowering plants. This comparison suggests how different the life forms in the two polar regions of the globe are.

Facing page:
Left: *Deschampsia antarctica*

Right: *Colobanthus quitensis*

Above left: A researcher studying the Algal bloom in the harsh environment.

Above right: *Prasiola calophylla*, a terrestrial green algae growing on the rock along with lichen.

Below: Microscopic snow algae growing in large quantities

Primitive life forms such as bacteria, fungi, algae, lichens and mosses are the only life forms that exist in some parts of Antarctica and usually these can be seen only during the Antarctic summer. *These are the only plants and microorganisms that provide food for most of the animals in Antarctica and they form an integral part of this fragile ecosystem.* During this period, the snow and ice gradually melt and expose only 2% of the landmass of Antarctica which supports the growth of these lower plants for a brief period. Beautiful patches of lichens, varying in colour from bright orange and yellow to green and black, can be seen on the surface of the rocks.

Algae are plant-like organisms that are the dominant photosynthetic life-forms in Antarctica. They include more than a thousand species of terrestrial, freshwater and marine forms (algae vary from microscopic single cell to massive seaweeds measuring several metres in length). Unlike plants they do not possess roots, stems and leaves. A representative of a group of algae called kelps, *Macrocystis pyrifera,* can grow up to seventy metres long. It attaches to rocks in coastal waters. Interestingly, the algae can survive in extremely low temperatures of -89°C in Antarctic as well as at high temperatures up to 70°C in hot springs. Algae in snow are found in beautiful and spectacular colours of red, green and yellow patches in the midst of brightly white snow. This is one of the most spectacular forms of photosynthetic life to be seen in the vast expanse of ice. Many varieties of terrestrial algae can also be found on exposed moist soil, in cracks and crevices in rock surfaces, on and below

Above: Walking on sea ice for sample collection. Dangerous. If you fall you are under the ocean! See footprints after the risky collection.

Below: Drilling the ice to collect the samples.

rock surfaces and on penguin guano. The greatest abundance and diversity of algae, ranging from blue-green to green algae and diatoms, can be seen in ponds, lakes and streams. During summer, these habitats receive water melting from snowfields and glaciers.

Above: Moss cushions growing adjacent to a stream

Below: Sahoo collecting Algae.

These algae can be found as thin and thick mats at the bottom and surface of the lakes. Besides this, a number of marine algal species can also be found inside the sea ice which are trapped during the formation of pack ice in the sea during the winter months.

Lichens of various colours growing on rocks

Collection of algae especially from the lakes and the sea is one of the most challenging jobs in Antarctica as one has to either enter the lake or the seashore. Getting inside these water bodies is tough due to the extremely low temperature of water. Another challenge is walking on the frozen lake and ice shelf. The green algae *Prasiola* grows very rapidly between December and March.

Lichens grow on rock surfaces with very attractive colours. Lichens have the ability to grow in extreme climatic conditions and thus some species are well-adapted to Antarctic conditions. There are a small number of moss species which form beautiful green cushions on exposed sandy and gravelly soils where there is sufficient water supply.

Animals in Antarctica

The largest land animals native to Antarctica are *not penguins or seals* as commonly perceived but is a wingless insect called *Belgica antarctica* which grows to a maximum length of 2-6mm. Antarctic land animals include fifty species of mites, twenty species of Collembola (Springtails), two species of midges, twenty-eight species of metazoa, twenty-eight species of tardigrades (minute water-dwelling invertebrates); several species of nematodes and protozoa. Springtails can survive temperatures as low as–50°C. One of the most challenging problems yet to be solved in Antarctica is the origin of these animals. Besides, there are several types of birds and other marine animals which are not native but inhabitants of Antarctica.

Belgica antarctica- the largest land animal of Antarctica measuring upto 6mm only.

Facing page: Emperor Penguins holding court. (Penguins and seals are not the native animals of Antarctica but they live together.)

However, Antarctica is free from mosquitoes that can occur in immense numbers in the Arctic.

Penguins

Penguins are one of the most human-friendly birds who welcome you to Antarctica. They are amongst the most beautiful groups of birds. Biologically, penguins belong to the order *Sphenisciformes* and family *Spheniscidae*. They are a group of aquatic, flightless birds living almost exclusively in the southern hemisphere, especially in Antarctica. Highly adapted for life in the water, penguins have contra-shaded dark and white plumage, and their wings have become flippers. Most penguins feed on krill, fish, squid, and other forms of sea life caught while swimming underwater. They spend about half of their lives on land and half in the oceans. Penguin biodiversity varies between 17-20 living species which is listed below with their biological names. Three most common penguins are Emperor penguin, King Penguin and Adelie Penguin. Details of the size, weight and lifespan of the various species are:

A large number of penguins dotting the coast.

Left and Left below (Emperor Penguins):
Emperor taking a cold bath (left); Emperor penguins politely bowing their heads (left below).

Right Below:
King Penguin
A king penguin awaiting visitors/ courtiers.

Emperor Penguin
Aptenodytes forsteri
Size: 100cm or 40 inches tall
Weight: 20 to 40kg or 45 to 90 pounds
Lifespan: 20 Years

King Penguin
Aptenodytes patagonicus
Size: about 90cm, 30 inches tall
Weight: 10 to 15kg, 20 to 30 pounds
Lifespan: 20 Years

Adelie Penguin
Pygoscelis adeliae
Size: about 60cm to 70cm (24-28 inches) tall
Weight: about 4.5kg (11 pounds)
Lifespan: 20 Years

Young chicks of Adelie penguin.

Right above: Penguins laying the egg.

Right below: An adult mother Adelie penguin trying to feed the young one.

Other species include:

Chinstrap Penguin	*Pygoscelis antarctica*
Fiordland Crested Penguin	*Eudyptes pachyrhynchus*
Erect-Crested Penguin	*Eudyptes sclateri*
Snares Island Penguin	*Eudyptes robustus*
Rockhopper Penguin	*Eudyptes chrysocome*
Macaroni Penguin	*Eudyptes chrysolophus*
Yellow-eyed Penguin	*Megadyptes antipodes*
Fairy (or Little, or Blue) Penguin	*Eudyptula minor*
African or Blackfoot Penguin	*Spheniscus demersus*
Humboldt Penguin	*Spheniscus humboldti*
Magellanic Penguin	*Spheniscus magellanicus*
Galapagos Penguin	*Spheniscus mendiculus*
Gentoo Penguin	*Pygoscelis papua*

Seals

Seals are *mammals* that live most parts of their lives on land and other parts entirely in the water. However, there are several Antarctic seals that very rarely pull out onto land. These are Weddell, Leopard, Ross and Crab-eater seals. Only Fur and Elephant seals breed on land in Antarctica. They look awkward on land, but are graceful in water. They give birth and care for their babies on land, but spend most of their lives in the ocean. Seals are related to walruses and sea lions, with which they are often confused. True seals, or earless seals, propel themselves through the water with rear flippers. They use side flippers to steer in the water and move about on land. Unlike sea lions, true seals cannot walk on their hind flippers when they are on land. Their torpedo shape allows them to swim quickly and efficiently. This shape also helps them escape predators, such as sharks, and catch prey, such as fish. Seals move so efficiently in the water that they can travel far from land to find food without using a large amount of energy. Seals can live in oceans as well as freshwater lakes, such as Lake Baikal in Russia.

However, most seals live in and around Antarctica and the Arctic Circle. Seals are well adapted to cold polar environments with thick blubber layers that act both as a food reserve and as insulation.

Six different species of seals live in Antarctic waters: *Ross, Weddell, Crab-eater, Leopard, Fur and Elephant seals*. The largest seal is the elephant seal. An elephant seal can weigh up to 4000 kilograms, which is about the weight of a minivan. The male elephant seal is two to three times larger than the female. Fur seals are the smallest, with adult females weighing only 150 kg.

Weddell's seal
Leptonychotes weddelli
Size: up to 3 metres
Weight: up to400kg

Crab-eater seal
Lobodon carcinophagus
Siz: up to 2.7 metres
Weight: over 250 kg

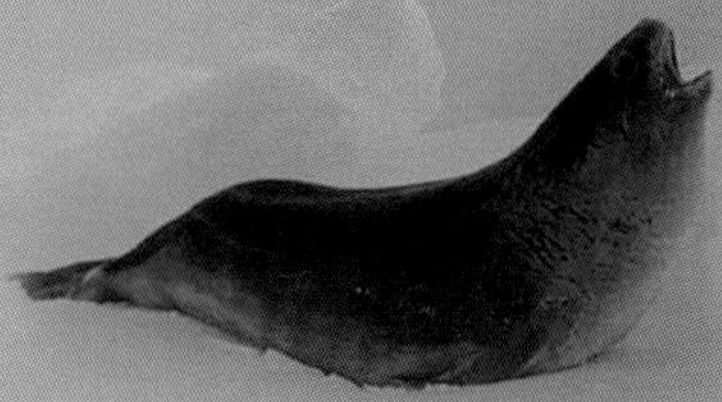

Ross seal
Ommatophoca rossi
Size: up to 2.3 metres
Weight: 200kg

Seals move so efficiently in the water, sometimes travelling on a piece of floating ice, that they can travel far from land to find food without using a large amount of energy. Seals can live in oceans as well as freshwater lakes. Their torpedo shape allows them to swim quickly and efficiently. Several Antarctic seals very rarely pull out onto land.

Leopard seal
Hydrurga leptonyx
Size: up to 3 metres
Weight: about 350 kg

Head of a Mummified seal. Antarctica is the best place for preservation.

Top: Leopard seal.

Right: Elephant seal.

Elephant seal
Mirounga leonine
Size: 6-7 metres
Weight: up to 4000 kg

Fur seal
Arctocephalus gazzela
Size: up to 2 metres
Weight: up to 150 kg

Krill

Antarctica has an abundance of a pelagic crustacean–the Antarctic Krill scientifically called *Euphausia superba*. It is a shrimp-like creature and forms a major food source for whales, seals, penguins, many fishes, squids and numerous birds. It is rich in protein and vitamin A. Some countries have already developed the capability to exploit and process this resource. Germany, Japan, Poland and Russia are major krill fishing nations today. These countries use krill meal as animal feed and are also experimenting with krill soup, krill paste, bread spread, krill fingers and sausages. Considering the abundance and easy availability of krill coupled with the depletion of the usual fishery stocks, it may well turn out to be a major fishery resource of the future. However, precaution must be taken not to overexploit this important resource.

How do plants and animals survive in Antarctica?

In general, when the temperature becomes extremely low, ice crystals are formed inside the cells which can damage the cells and ultimately cause death. There are many such animals and plants in and around Antarctica which have the presence of some anti-freezing molecules inside their cells which do not allow the formation of ice crystals inside the cells or freezing of the cells thus preventing death of these valuable lives.

Left (above and below): Antarctic krill. Food for future.

Origin of Antarctica

The Greek opinion

It was the ancient Greeks who first came up with the idea of Antarctica. *They knew about the Arctic–named Arktos*–The Bear, from the constellation the great bear and decided that in order to balance the world, there should be a similar cold Southern Landmass that was the same but the opposite "Ant-Arktos"–opposite The Bear. They never actually went there. It was just a lucky guess or a statement based on pure logic!

As early as the 6th century BC Pythagoras postulated that the world was round. Parmenides (450 BC) and Aristotle (384-322 BC) agreed to this. Once it was accepted that the earth was a sphere, there came into being many theories about its nature. Parmenides held the view that the globe was divided into five parallel climatic zones. The existence and size of southern continent (today's Antarctica) was a matter of intense interest even during the earliest civilisations, with the early Greek philosophers propounding the need for a southern landmass to balance the weight of the northern hemisphere. And as the natural hemisphere lay under the constellation of Arktos, the Bear, so Aristotle (384-322 BC) reasoned, the unknown land to the south could be Antarktikos–in other words the total opposite.

Geological history and palaeontology

More than 170 million years ago, Antarctica was part of the supercontinent Gondwanaland. Over time,

"Ant-Arktos": The wide white expanse that is Antarctica.

Gondwana gradually broke apart and Antarctica as we know it today was formed around twenty-five million years ago. During the Cambrian period, Gondwana had a mild climate. Towards the end of the Permian period, continued warming led to a dry, hot climate over much of Gondwana. As a result of continued warming, the polar ice caps melted and much of Gondwana became a desert. In Eastern Antarctica, the seed fern became established, and large amounts of sandstone and shale were laid down at this time. Synapsids, commonly known as "mammal-like reptiles", were common in Antarctica during the Late Permian and Early Triassic and included forms such as *Lystrosaurus*. The Antarctic Peninsula began to form during the Jurassic period (206–146 Ma), and islands gradually rose out of the ocean. Ginkgo trees and cycads were plentiful during this period. In West Antarctica, coniferous forests dominated through the entire Cretaceous period (146–65 Ma), though southern beech began to take over at the end of this period. Ammonites were common in the seas around Antarctica, and *dinosaurs were also present,* though only three Antarctic dinosaur genera (*Cryolophosaurus* and *Glacialisaurus*, from the Hanson Formation and *Antarctopelta*) have been described to date. It was during this period that Gondwanaland began to break up.

The cooling of Antarctica occurred stepwise, as the continental spread changed the oceanic currents from longitudinal equator-to-pole temperature-equalising currents to latitudinal currents that preserved and accentuated latitude temperature differences.

Break up of Gondwanaland and movement of continents

Africa separated from Antarctica around 160 Million years ago, followed by the Indian subcontinent, in the early Cretaceous (about 125 Million years ago) period. About sixty-five Million years ago, Antarctica (then connected to Australia) still had a tropical to subtropical climate, complete with a marsupial fauna. About 40 Million years ago Australia-Papua New Guinea separated from Antarctica, so that latitudinal currents could isolate Antarctica from Australia, and the first ice began to appear. During the Eocene-Oligocene extinction event about thirty-four Million years ago, carbon-di-oxide levels have been found to be about 760 ppm and had been decreasing from earlier levels measuring in thousands of ppm. Around twenty-three Million years ago, the Drake Passage opened between Antarctica and South America, resulting in the Antarctic Circumpolar Current that completely isolated the continent. Models of the changes suggest that declining carbon-dioxide levels became more important. The ice began to spread, replacing the forests that then covered the continent. Since about fifteen Million years ago, the continent has been mostly covered with ice, with the Antarctic ice-cap reaching its present extension around six Million years ago.

Exploring the continent

Captain *James Cook*, a British navigator sailed from England in July 1772 in an expedition in the search of the southern continent. On January 17, 1773 he and his team became the First to cross the Antarctic Circle (67°15' South) but could not proceed further as the sea was frozen with ice along the surface layers. He was just eighty miles away from the mainland Antarctica. Although Cook crossed the Antarctic Circle thrice between 1772-1776, he never claimed to have seen the southern continent.

Till 1819, no human being had seen Antarctica. The *first person* to see Antarctica was Captain Thaddeus Von Bellingshausen, a Russian explorer who saw the continent from a distance on January 27, 1820 but could not confirm this discovery. The *first human landing* on the continent was by John Davis, an American sailor, who probably landed on Antarctica on February 7, 1821 (this landing is not certain). From this time till January 1895 for nearly seventy-four years the mainland Antarctica remained almost unexplored although there were several activities along its periphery.

Some Milestones in Early Explorations

February 20, 1823: James Weddell, a British explorer reached 74°15' S of the Antarctic Circle around 345 km further south than James Cook had achieved by reaching 67°15' S on January 17, 1773. (Weddell Sea has been named after James Weddell).

April 10, 1832: John Biscoe, a British explorer, claimed that, "I am firmly of the opinion that this is a large continent," and "I saw to an extent of 300 miles (482 km)". He completed one of the first circumnavigations of the Antarctic continent.

The Southern magnetic pole is slowly drifting northwards. The earth's magnetic poles are the points where a compass needle will stand vertically.

1840: The First attempt to reach the south magnetic pole was made by Dumont d' Ureville (a Frenchman) and Charles Wilkes (an American) to reach 66°S and 144°E. Wilkes was the first to call it the *Antarctic Continent.*

March 5, 1843: Captain James Clark Ross (a British Naval officer in the Arctic navigation) crossed the Antarctic Circle three times (1839-1843). Ross was a unique person, having spent four and a half years in the Arctic during 1829-1833. He located the position of the North Magnetic Pole in Northern Canada and subsequently was entrusted with the responsibility to locate the Southern Magnetic pole. He was able to determine that the Southern Magnetic pole must be approximately at 75°30′ S and 154° E in Victoria Land but could not locate the actual site. The South Magnetic pole (72°24′S, 155°18′ E nearly 370 km north of the site indicated by Ross) was finally reached by Douglas Mawson, E. David and A. Mackay in 1909 through a tough journey carried out using the sledge.

Today the south magnetic pole is actually 1,800 miles (2,900 km) away from the earth's true south pole whereas the north magnetic pole is fairly near to the true north pole and is 'only' about 360 miles (580 km) away.

In Ross's expedition, a twenty-two-year-old boy called Joseph Dalton Hooker sailed to Antarctica. Subsequently, Hooker became a great naturalist and nineteenth century's most famous scientist. He was later appointed as the Director of the famous Kew Gardens in 1865. He died in 1911.

January 24, 1895: The first confirmed landing made on the continental mainland Antarctica was on this date. This expedition was launched by a Norwegian businessman named Henryk Johan Bull who left Tønsberg, Norway on September 20, 1893 in an old 226 tonnes renamed ship called *Antarctic* with thirty-one crew members.

By the end of the century every Antarctic explorer wanted to set his foot first on the South Pole and be a part of history. Thus began the journey to the extreme point. Captain Robert Falcon Scott, a Britisher was the first person to accept this challenge.

1901-1904 Robert Scott's Discovery Expedition:

After months of preparation Scott, with his team which also included Edward Wilson and Ernest Shackleton, started his journey to Antarctica on August 6, 1901 in a ship called *Discovery* and reached the Ross Ice Self on January 23, 1902. On February 4, 1902, Scott became one of the first "Aeronauts" to have the first ever balloon ascent over Antarctica. Between 1901-04 his team made several attempts to reach the South Pole but could not make it due to several difficulties such as extreme blizzard conditions, frostbite, lack of sufficient food, etc. A few team members as well as the dog pulling the sledge died on the way. Several members suffered from Scurvy (a disease caused due to Vitamin-C deficiency). Thus the expedition was abruptly called off when they had reached 82° 17′ S while still 850 km away from the South Pole.

1907-1909 Ernest Henry Shackleton's "Nimord Expedition":

This first experience and personal failure in Scott's "Discovery Expedition" between 1901-04 determined his return to Antarctica in 1907 as the leader of the *Nimord Expedition*. On January 9, 1909, Shackleton, with his colleagues, could reach up to 88° 23′ S which was 180 km from South Pole.

Shackleton's Hut.

1910-1912 Roald Amundsen's successful expedition to South Pole

Roald E. G Amundsen, a Norwegian explorer, first made plans to visit the North Polar Basin, but due to several reasons he changed his journey towards Antarctica. Drawing lessons from the failure of others, Amundsen meticulously planned his South Pole trip. He chose the best of clothing, equipment, food, dogs, etc. including the choice of team members selected to go with him. Amundsen and his team left Oslo, Norway on June 3, 2010 and arrived at the

The key to Amundsen's success was his use of the dogs. He was both practical and unsentimental about them. He chose them from North Greenland as they were used to polar conditions. He picked up 97 dogs to pull the sledges and assist them. At the same time he also planned to kill some of his dogs on the way so that he could use them as a source of fresh meat. Thus, when he reached the South Pole only 16 dogs from the original pack of 97 were left.

Above: Good food, equipment and other supplies were vital during the expedition.

Below: Dress-up for sledging with complete protection.
Prof Paul Broady.

eastern edge of the Ross Ice Self on January 14, 1911. With remarkable courage Amundsen, with his five-member team, finally reached 90° S the geographic South Pole at 3 O'clock on Friday December 14, 1911, creating history. He raised the Norwegian Flag and kept the records of reaching the South Pole in a safe box. Finally, the team returned to Hobart, Australia on March 7, 1912.

Scott's second attempt to South Pole

Failure to reach South Pole in his first attempt between 1901-1904, Scott announced his intention to lead the expedition again to Antarctica. The ship *Terra Nova* left Britain on June 1, 1910 but Scott joined the ship only at South Africa. While the ship was passing through Melbourne, Australia, Scott received a telegram from Amundsen that his Norwegian team was also on its way to the South Pole. Thus began the race between the two legends as to who would make it first. On January 9, 1912 Scott with four other members passed Shackleton's furthest South record of 88° 23' S. Finally, after many a struggle he reached the South Pole on January 17, 1912 i.e. 33 days after Amundsen's arrival. Seeing the Norwegian flag Scott's anguish was unrestrained: He wrote: *"The Pole. Yes, but under very different circumstances from those expected... Great God! This is an awful place and terrible enough for us to have laboured to it without the reward of priority."* He then added: *"Now run for the home and desperate struggle. I wonder if we can do it."* The whole team was depressed and they had to trek back 1290 km. They were facing shortage of food, feeling weak and depressed and two of the members had already died during February.

Above: Dogs were used in the olden days for expeditions.

Scott's party at the South Pole, 18 January 1912. L to R: (standing) Wilson, Scott, Oates; (seated) Bowers, Edgar Evans.

Right: The routes to the South Pole taken by Scott (green) and Amundsen (red), 1911–1912.

Below: New Amundsen-Scott station at south pole is an engineering marvel.

By March 22, 1912 the weather had worsened and their fuel was exhausted. On March 29, 1912 Scott finally wrote in his diary: *"We are getting weaker and the end cannot be far. It seems a pity, but I do not think I can write more. For God's sake look after our families."* The frozen bodies of Scott and two other team members Wilson and Bowers were found eight months later by a search party. The death of Scott and his party made headlines around the world. Amundsen noted: *"I cannot read that last message of Scott's without emotion... And to think that while those brave men were dying out there in the waste of ice, I was lecturing in the warmth and comfort in Australia."* (Amundsen returned to Australia from South Pole on March 7, 1912). Besides the King of Britain, the President of the United States of America, William Taft, mourned the death of Scott which had a worldwide impact. *The American research station at South Pole was thus named as Amundsen-Scott station in their memory.*

Subsequently, nationalistic motives, personal ambition, love for adventure, thirst for knowledge, desire for commercial exploitation and many other factors attracted human beings for the exploration of the continent.

Geographic South Pole at 90°s.

Left: A US Hercules Aircraft used for Modern Day's transportation.

Right above: The research station of USA at south pole popularly known as Amundsen-scott station. This station is no more in use.

More than 98% of Antarctica is covered with ice. It is the driest of all continents where nearly 90% of the world's fresh water is stored in the form of ice. Surrounding the continent is the Southern Ocean which is frozen into ice during the Antarctic winter.

India's Journey to Antarctica

India's Journey to Antarctica

A first person account by Dr S.Z. Qasim.

Dr S.Z. Qasim, the leader of the First Indian Scientific Expedition to Antarctica in his own words:

January 9, 1982: Indian team led by Dr S.Z. Qasim landed on Antarctica. The inspiring spirit behind the First Indian Scientific Expedition to Antarctica was the former Prime Minister of India, the late Mrs Indira Gandhi, whose initiative and interest served as a guiding force.

Origin of Indian Antarctic Expedition:

During the 1980s, when the idea of launching an expedition to Antarctica was being considered, there was a lot of scepticism and uncertainty about the venture. I was then the Director of National Institute of Oceanography, India's leading Oceanographic institute, at Goa. I was on a historic cruise as the Chief Scientist on Board from January 17, 1981 until February 5, 1981 for exploration of polymetallic nodules in the Indian Ocean region. The First sample of polymetallic nodules was hauled on January 26, 1981. Subsequently massive deposits of these nodules which contain copper, nickel and cobalt, besides manganese and iron ore was located in the western Indian Ocean. With this breakthrough, India became the first country in the Third World and the seventh in the world to enter the field of deep sea exploration of minerals. Thus India had gained recognition as one of the pioneer investors for seabed mining.

Dr Qasim sitting on the ski-doo, the snow scooter.

Facing page: *Polar Circle* leaving Indian coast and moving towards Antarctica.

It was mid May 1981 and I had just joined as the first Secretary of the newly formed Department of Environment (now Ministry of Environment and Forests), Government of India, at New Delhi. As it is mandatory to officially call on the Cabinet Secretary of India after joining, I met C.R. Krishnaswamy Rao Sahib, the Cabinet Secretary, in his office at South Block. During the meeting he casually asked me that since I have so much experience about oceans, whether it is possible for India to visit Antarctica. My reply to this question was in the affirmative and the dialogue ended there. Subsequently, the Cabinet Secretary, while meeting the then Prime Minister Mrs Indira Gandhi, mentioned to her that the new Secretary of Environment, Dr S.Z. Qasim, who has recently joined his post in Delhi, is very confident about India's proposed visit to Antarctica. The reason for this was that the proposal on India's visit to Antarctica had been under consideration at that time by the Government and my comment, in a light-hearted manner, was taken seriously. The Prime Minister expressed her desire to meet me and discuss about the possibilities of an Indian expedition to the Antarctic. Soon afterwards I met Mrs Gandhi along with the Cabinet Secretary and she asked about the feasibility of India's expedition to Antarctica. My answer was almost the same as before. She wanted to know what the essential requirements would be if we were to make a maiden trip. I mentioned that the foremost requirement would be the chartering of a suitable ice-breaker ship which had been to Antarctica before. It needed to have an ice-strengthened double hull for safety in case of the ship accidentally hitting a sharp ice boulder. The other major requirement was that the Captain of the ship should have had the prior experience of having visited Antarctica earlier and had thus the experience of navigation in icy conditions.

Left: Mrs Indira Gandhi, the then Prime Minister of India.

Right: Members of the First Indian Expedition along with their crew members after their departure from Goa.

Journey of the Leader: The final decision to launch an expedition was taken by the Prime Minister in July 1981 and the operation was code named *"Operation Gangotri"*. The Council of Scientific and Industrial Research

(CSIR) was selected as the lead organisation with the National Institute of Oceanography as the coordinating agency for the First expedition. The 21-member team belonging to seven institutions included thirteen scientists who were carefully selected by top ranking scientists of the country and eight Logistics personnel drawn from Indian Defence Forces. After a rigorous medical examination and psychological test, the team underwent a training camp in a cold region of the Himalayas for acclimatisation in various aspects. Finally on *December 6, 1981*, the Indian team sailed with twelve crew members from Goa, in the ice-breaker vessel named *M.V. Polar Circle* chartered from Norway. One of the biggest challenges for the crew was to cross the Roaring Forties that lies between 40-50 degree South latitude. The sea was extremely rough and dangerous with high waves. It was impossible for the members to eat anything and sleep as this was the first such experience for all the members of the team. After crossing this obstacle, the next hurdle was crossing the pack ice at 55 degree South. Huge masses of ice were formed due to the freezing of seawater. As the ship proceeded towards the 60 degree South mark towards the Southern Hemisphere, the thickness of the ice increased forming ridges at many places. Often the ship got stuck in thick ice and at such moments water had to be pumped from the rear to the front to melt the ice barrier by application of heat. For seven consecutive days, the ship was breaking the ice and the weather condition was extremely bad with very poor visibility. The thickness of ice was so huge that it was difficult to recognise the main continent and we lost our way. The hostile weather, loss of sleep, deep psychological pressure, prolonged isolation from civilisation and the fear of death was prevailing upon the minds of the members. Everybody had lost hope of landing in Antarctica, forget about his return to India. But as the leader of the team, I did not lose hope. Many times, I used to boost the morale of my team by saying, "*Pals I am not bothered about my own safety. I am more concerned about you...death can come in a road accident, in sleep, or in various other ways but there can be no better death than one on this trip, for our country. If we die now there will be the entire nation of 68 crores of people to mourn us*". I inspired all my team members to be fearless and pursue the mission with confidence and optimism. In the meantime we had made three attempts to land but could not. When everybody decided to come back without landing, I told the members you may go back but I will not. At one time I decided to take the helicopter from the ship and complete my mission. I told all my members, "What face shall we show to the people of India, if we fail in this mission! The country's prestige will go down in the international community." With luck on our side the weather changed dramatically and we could see a Japanese Ship *Fuji*. On January 8, we regained some hope and allowed the ship to move forward. Thus in our fourth attempt we decided that the helicopter should guide us from above showing the route to reach the coast. Finally with this last attempt the expedition landed on Antarctica on January 9, 1982 at 3:00 am Indian Standard Time. As a

Dr S.Z. Qasim, the leader of the First Indian Scientific Expedition to Antarctica hoisting the Tricolor on January 9, 1982 after landing.

A

B

D

F

A: Camping site for the first expedition

B: Engaged in scientific activities during the first expedition

C: The ship *Polar Circle* is anchored close to the camping site. The naval helicopter is being covered to protect it from heavy blizzard.

D: Dr Qasim watching the penguins.

C

E

G

E: Leading from the front to explore the continent.

F: Taking rest after a long day of journey.

G: A hero's welcome to the first triumphant team.

mark of victory for the nation I along with my team members hoisted the Indian Flag on Antarctica on January 9, 1982 thus laying the foundation for Polar science for India. On landing in Antarctica, the first most exhilarating experience was to see the local animals–the penguins and the seals. We quickly established the communication link to India and I spoke to the Prime Minister giving her the good news of our successful landing. The Prime Minister said, "*What thrilling news that our Indian Expedition has landed in Antarctica. It is the fulfilment of one of my long standing wishes... The entire area is of deep interest to us and of vital importance. This will take us forward to greater success in science.*"

Above: The Prime Minister honouring Dr S.Z. Qasim in New Delhi.

Below: Dr Qasim introducing the members of the expedition to Mrs Indira Gandhi.

The team built the base camp and carried out the necessary exploration and experiments successfully. Finally we left Antarctica and the expedition, after covering a distance of 21000 kilometres returned to Goa on February 21, 1982. The duration of the expedition was seventy-seven days.

Dr Qasim being congratulated by the other Indian scientists after a successful expedition.

After the successful completion of the expedition, Mrs Indira Gandhi made the following comments: "The expedition gave us the opportunity to work at sea and on the icy continent, in most unpredictable and hostile weather conditions. The number of tasks completed and the amount of scientific data generated clearly shows that with dedication and pioneering spirit, Indian scientists can accomplish more than what is initially expected. It also proved India's capability to undertake Antarctic exploration of a high order..."

The late Prime Minister met each member of the expedition team, participated in a group photograph and gave beautiful mementoes to the members.

The first expedition thus laid the foundation of Polar research in India. It collected important scientific data on meteorology and hydrology, installed an automatic weather recording station, made use of navigational equipment and position-fixing, studied physical characteristics of the Antarctic convergence, carried out experiments on radio-wave transmission, made chemical studies of the self waters, ice self, freshwater lakes, distribution of trace metals in south-western Indian Ocean, aerosol particles, sedimentology of the seabed, characteristics of the rock samples, geochemistry of the metapelite, biological productivity of self waters, zooplankton, biomass, benthic fauna, sea birds, marine mammals, etc. In addition, it also surveyed the ice self region and the Queen Maud Land area for the site of a permanent station. According to the best traditions of the first visiting team to Antarctica, it erected a refuge hut and packed it with food, fuel, medicines, etc., to be used by any other party in future. Some members of the scientific team also visited the neighbouring USSR station *Novolazarevskaya* and exchanged both logistic and scientific information. One of the important achievements of this expedition was the installation of an automatic weather station to record continuously on a cassette recorder, the weather data of Antarctica for a period of one year. The station was erected in the Schirmacher Hill region. It worked continuously for almost a year till the cassette recorder lasted. When the team of the second expedition reached the site, it retrieved the cassette. Other important landmarks during the first expedition were the discovery of an undersea mount which was named as *Indira Mount,* and geomagnetic intensity recorded during the period of stay which occasionally showed the occurrence of magnetic storms.

The main objectives of the First Indian Antarctic Expedition were as follows:

1. To initiate studies, build facilities and carry out observations on different oceanographic disciplines.
2. To continue and strengthen programme of continuous data collections and studies.
3. To identify programmes of significance to the Indian context in scientific and economic terms and pursue these as thrust areas to establish a position for Indian science in this sector.

Establishment of *Dakshin Gangotri*: The necessity of having a permanent base station where scientific observations and research could be continued during the whole year including the winter months was felt strongly. Thus the Third Indian Expedition comprising eighty-one members was flagged off on December 3,1983 carrying a large amount of prefabricated building material in an ice breaker ship *Finnish Polaris* which was leased from Finland.

Thus began the construction of India's first permanent station. The station consisted of two storeyed blocks linked by a narrow passage. The building was made up of timber, prefabricated panels which were filled by insulated material. The outside face of the structure was protected by a thick layer of insulated material. The building had two blocks in which the first block housed essential services including generators, fuel-supply-systems, workshop, storage space, etc. while the other block included a research lab, kitchen, living room, toilet, etc. The station was supplied with power supply for twenty-four hours through three large generators. There was a well organised central heating system to maintain the inside temperature at 15° Celsius. For entertainment there was a music system, VCR, table tennis, gymnasium, and other indoor facilities.

The most important milestone in Indian Polar science was laid when the first permanent station *Dakshin Gangotri* (DG) was built on the ice self (latitude 70° 05′ S and longitude 12° 00′ E) during the course of the third expedition. This expedition also included the *first* wintering team of twelve members. A permanent communication system was established during the third expedition. A detailed programme of biological oceanography was started during this expedition. The country was consolidating the achievements of

Right: Construction in mid way.

Laying the foundation for construction of *Dakshin Gangotri* on December 29, 1983.

the first two expeditions and planning to launch the future ones, thereby confirming its sustained interest and proven capabilities to conduct front-ranking Antarctic Science research. It was admitted to the Antarctic Treaty System on August 19, 1983 and soon thereafter, was granted a Consultative Status on September 12, 1983. On October 1, 1984, India was also admitted to Scientific Committee on Antarctic Research (SCAR) and later became a member of Convention on the Conservation of Antarctic Marine Living Resources (CCAMLR) in 1986.

Celebrating India's Independence Day on August 15, 1984 at *Dakshin Gangotri.*

Above: Painting the roof in Indian Tricolour on January 16, 1984.

India's second permanent research station *Maitri* was built at Schiramacher Oasis.

Dakshin Gangotri (DG) was decommissioned in 1990 and converted into a supply base. The expertise and knowledge gained in low temperature structural engineering from the first station, *Dakshin Gangotri* was applied in setting up the second Indian station, *Maitri*, entirely through indigenous efforts.

Establishment of *Maitri*: Second Indian Permanent Station: As is true of any material left on ice, because it slowly sinks down, so did the first station *Dakshin Gangotri*. By the time its warranty period for safe living was over, a new station on the rocky hill ranges named as Schirmacher Ranges, less than 100 km further from south, was established and named *Maitri*. The construction of this station started during the seventh expedition (1987-88) and it was

completed during the eighth expedition (1988-89). One of the significant features of this expedition was the participation of a young research scholar, Dinabandhu Sahoo (now a faculty member of Delhi University and the co-author of this publication). He was the First ever Indian Student to participate in the Indian Scientific Expedition to Antarctica.

Journey of the First ever Indian Student to Antarctica

Journey of the First ever Indian Student to Antarctica

A first person account by Dr Dinabandhu Sahoo

After finishing my Master's Degree in Botany from the University of Delhi, Delhi, India, I was registered as a PhD student to pursue my research in the field of marine algae, more commonly known as seaweeds. Since I am from the coastal town of Puri located in the state of Odisha, I was always fascinated with the ocean and wanted to explore it. The desire was always to learn through exploration and adventure. So, I used to backpack travel to different remote parts of the Indian coastline for my research material collection. It was early 1987 and one day, I saw an advertisement regarding selection of scientists for participation in the Indian Scientific Expedition to Antarctica. Then a number of thoughts rushed across my mind: Will it be possible for me to take part? How would I get there? Will the Government of India select me? Can I compete with other scientists since I am only a student? Will the University forward my application? Those were the questions which haunted my mind for which I had no answers. One fine morning I picked up my scooter and drove straight to the Department of Ocean Development (Now, Ministry of Earth Sciences) to enquire and know more about the expeditions. In the Ministry, the officer dealing with this proposal told me that it is a very tough competition. He advised me to forget about participation as this was for very high-level, reputed scientists. He said, you are just a student and we do not have any provision for any student in such an important activity which is of national importance, the government invested a lot of money in this. I came

Above: Dinabandhu Sahoo driving a snow scooter.

Facing page: Sahoo hoisting Delhi University flag at *Dakshin Gangotri* Antarctica on 26th January 1988.

back highly demoralised that afternoon and returned to my laboratory but could not concentrate on anything.

The next day I prepared a project and showed it to my supervisor Late Prof. M.R. Vijyaraghavan, who liked the idea and encouraged me to apply for the expedition. With my supervisor's blessings and best wishes I went ahead and submitted my application. There were several other applicants like me from other parts of the country. Days, weeks and months passed. Each day I spent was with a lot of hope, anxiety and tension and at the same time with curiosity and apprehension. One afternoon I received a telephonic message through my department. that said, "You are urgently required to meet the Secretary, Department of Ocean Development, Government of India."

With a lot of fear in my mind, very natural in any student's mind, I visited the Secretary's office. It was a delightful moment in my life to meet the leader of the First Indian Scientific Expedition to Antarctica and the Secretary, Department of Ocean Development, Government of India, Dr S Z Qasim. He offered me a cup of tea and made me very comfortable, although I was very nervous and sitting stiff like a statue. After several minutes of interesting discussions, he told me he had good news and that I had been selected as the First Indian Student to take part in the forthcoming Seventh Indian Antarctic Expedition. In the wildest of my imagination or dreams I had not thought of this happening. I could not believe it and remained dumb for a few seconds. This was a very happy moment in my life, no, in fact the happiest moment in my life. After a few days I got an official letter from the ministry to undergo medical examination and training at High Altitude Warfare School.

Glacier Training at Himalayas before Expedition:

After successfully passing the medical fitness test I was advised to join other selected team members to undergo an extensive familiarisation and acclimatisation training at Indian Army's High Altitude Warfare School (HAWS) located near Sonmarg in the state of Jammu and Kashmir. It was a great experience to undergo various types of training such as mountaineering, river crossing, rock climbing, glacier walking, ice wall climbing, crevasse crossing and rescue, etc. The most interesting aspect was to have "Igloo Training". We were taught how to survive in the snow and blizzard. In case of eventualities how one could dig the ice, make ice-

Icewall climbing.

Facing page (Both):
Training at High Altitude Warfare School, Himalayas.

bricks, join the ice bricks to make an igloo and stay inside. Even if one lights a candle inside the igloo it can provide a lot of warmth. Undergoing this type of training, the first of its kind in my life was a unique experience. In my opinion everybody should undergo such training which not only raises the level of self confidence but also helps in future. This ten days short but rigorous training from September 21-30, 1987 was a great experience and still remains fresh in my memory. Although, our station commander was a very tough man and a hard taskmaster but at the end it was very useful.

Preparation:

After successfully completing the training at HAWS, I received my final selection letter to participate in the expedition. The letter clearly explained the dos and don'ts. The most important challenge was to pack up the baggage from pins to microscope to equipments, chemicals, books, pen, paper, etc as well as every possible little thing that you could think of. The biggest challenge for me that time was to buy a good quality camera with high power lenses. Those days, in 1980s, it was very difficult to buy a good camera as these were not available and the price was too high for a student to afford. Since, I was getting a few hundred rupees in monthly stipend and scholarships per month; I had a meagre savings ₹3000 to meet the expenses of all the purchases I needed to make for the trip. But a good camera with high power lenses used to cost around ₹20,000 in those days. I had no idea what to do. Without a camera, how could I manage? This was a once-in-a-lifetime opportunity and I felt I should somehow ensure that I had a camera. One day I told my Japanese friends Shunji Hosaka and Hirotaka Sato who were studying at Delhi University that I would be going to Antarctica, shortly, but they could not believe it. They were very excited and offered to render any type of help. They asked me to let them know if I needed anything from Japan. With lot of hesitation I asked them if they could buy a good quality camera for me, I will bear the cost, I assured them. They replied that they would try. After few days, they brought a wonderful camera for me at a discounted price. Although it was very expensive, they asked me to pay them just ₹8000. I was delighted as one of my major problems was solved. Probably, without their timely help I could not have got such a good camera nor taken the beautiful pictures you will come across in this book.

A few days before my departure to Antarctica the then Vice-Chancellor, late Professor Moonis Raza, invited me to his office for a cup of tea. He told me that the whole University was proud of me, especially the student community. He, on behalf of thousands of students, teachers and employees, conveyed the best wishes and good luck for my journey, *Bon Voyage*. With these words, he handed over to me the Delhi University flag to hoist in Antarctica. I must confess here that he was one of the most student-friendly Vice-Chancellors I have ever met. He was a wonderful orator who used to encourage and enthuse students with his charming words. Finally, with all these good wishes from different quarters I boarded the train from New Delhi railway station to Goa with two heavy trunk-loads of necessary equipment, chemicals and my backpack. Just before our departure from Goa to Antarctica we had a wonderful briefing party where all the expedition members, ministry officials, some previous expedition members, ministers, VIPs and crew members of the ship were present. Dr Qasim met each member, briefed them and wished them good luck and success.

Departure from India:

Farewell ceremony: I, along with other team members from various participating organisations and a large logistics team drawn from the Indian Army, Navy and Air Force, was given a grand ceremonial send off at Marmagao Port, Goa. A large number of family members, defence personnel and friends were gathered at the port to bid farewell to us. The atmosphere was quite emotionally-charged. The military band was playing the tune and drumming the beats of "*Saare Jahan Se Achcha, Hindustan Hamara*". While the drumbeats were resounding with full gusto, the proud feeling of being an Indian was deeply penetrating my mind and heart, making me emotionally surcharged. With full military honour, to the echo of drumbeats, we slowly climbed up the ladder step by step and finally lined up on the deck of the ice-breaker ship *MS Thuleland* chartered from Sweden. Against the background of the setting sun, our huge ship finally started moving away from the Indian shores towards our final destination. As we slowly drifted away tears of happiness were rolling down my cheeks with fading drum sound from the distance. Finally, our ship entered into the huge vastness of the Arabian sea with complete darkness. We left India on November 25, 1987.

About the Ship: An ordinary ship cannot go to Antarctica as the sea gets frozen into very thick ice which needs to be broken to create a passage. That is the reason our ship *MS Thuleland* was a huge ice-breaker ship chartered from Sweden. The ship was nearly 186 metres long and 27 metres wide weighing 9,686 metric tons and having a capacity of 31,900 metric tons. This ship was built in 1977 and consumed 20 tons of oil per day for running. The ship had six holds to carry the cargo and supplies. The ship had well equipped air conditioned cabins for housing nearly hundred members, a well equipped kitchen, gymnasium, sauna, laundry, a small library, table tennis games facilities and two helipads. One part of the ship was modified for various entertainment activities with a huge hall that could accommodate hundred people at a time, where people could meet and relax, or watch television and videos. The ship carried thousands of barrels of diesel as supply material for power generators, snow scooters (skidoos), snow trucks. It also had huge stocks of various types of

The 186 metre long icebreaker *MS Thuleland*.

frozen and dry food material, besides, biscuits, ice cream, fruits, vegetables, clothes, magazines, books, audio and video cassettes and many other such items of everyday needs.

The journey in the sea: The captain of the ship along with other crew members welcomed us on board. The next morning , the leader of the expedition and the ship's captain addressed the members regarding the safety measures to be taken during the journey as well as in Antarctica. We were given the life boat demonstration and taught survival lessons in the sea. Each member, irrespective of their designation or position was assigned a daily duty to be performed in the ship's kitchen on rotational basis which included cleaning the kitchen, washing utensils, laying the table, serving food, cleaning the hall, etc. Travelling in the sea and the solitude it provided was fun for the first couple of days as I could see only the deep blue water around me in the vastness of the great Indian Ocean during the day and complete darkness in the night with the sound of the waves as company. As we approached the 0° Latitude (*The Equator*), preparations for "*Equator Crossing Ceremony*" began. On the Fifth Day when we reached the Equator, the ship stopped for a few hours. All the members came out on the deck for a grand celebration. One of the crew members sat on a throne which was raised on the deck, wearing a big crown on his head posing as the *King Neptune*. Besides him there were four other persons who were dressed as soldiers holding scissors, swords, sticks, etc. It resembled a maharaja's court. Each team member was presented before the king, one by one , for certain rituals which included a final holy dip inside a small artificial pool raised on the open deck of the ship. It was a great festive atmosphere with a lot of fun in the middle of the sea. The basic idea of the whole ceremony was to pray to God for a safe passage and safer return.

On the Tenth Day we passed the coast of Mauritius (the last landmass we saw on our journey) and the next day we crossed *Tropic of Capricorn* (23.5^{0} South Latitude). Various types of woollen clothes, caps, gloves, life saving suits, ice axe, snow boots, sunglasses, etc were distributed to the members. Members were also given their quotas of dry food, snacks, juices and cold drinks to consume during the expedition.

Rough sea and Life in Roaring Forties: Thirteen days had passed and suddenly we found the ship was rolling and pitching making life uncomfortable. I came out of my cabin to the deck and found that the wind was blowing at a very high speed and the speed of the ship had been reduced to only a few knots (1 knot = 1.852 km). On the Fourteenth Day, huge waves ranging from 15-20 ft in height were splashing all around the ship with a roaring sound. I was standing on the deck watching the fury of the ocean when suddenly the ship started pitching and high waves lashed across the deck of the ship. I fell on the deck and thought that the ship was going to capsize or be submerged in the middle of the sea and that all of us were going to die. With strong fear and nervousness taking a firm grip on my mind, I couldn't figure out as to what was happening, I was numb for a moment.

I got up with utmost difficulty and rushed towards the door of the main hall. The wind pressure was so high from the outside that it was extremely difficult to open the door. At the same time the roaring sound of the wind and the lashing of the waves were so strong it was difficult to stand outside for even a minute. Standing alone on the deck in that situation was very scary, I was praying to God as how to open the door and rush into my cabin and save my life. I felt a chill running down my spine. In the meantime, I was completely drenched. Anyhow, after a lot of struggle and fervent prayers, finally God listened, took pity on me and I was able to get inside my cabin to my great relief. When I asked the other team members why the sea was so rough, they replied we were passing through the "Roaring Forties".

Although the whole incident was very scary and highly traumatic, I regained my composure and felt that I had to capture these moments in my camera as this was a once-in-a-lifetime opportunity. My heart thudding against my ribs, I immediately changed out of my wet clothes, picked up my camera and then rushed to the pilot's cabin to capture glimpses of these rare moments. This rolling and pitching continued in the high seas for nearly 2-3 days. Many of the members suffered from sea sickness with vomiting, nausea and dizziness being the most common complaints. No amount of medicine worked to cure the sea sickness. Walking, eating, sleeping, bathing and

Ship passing through the most turbulent ocean (Roaring Forties).

Many a times the ship would sway to such an extent that, while seated in the dining hall one could see the ocean from one side of the window and the sky from the opposite side tilting the whole body into imbalance in a difficult angle. I would like to mention here that our ship's rolling capacity was up to 35° and during the journey it rolled between 25-28°. Since this was the most turbulent period during our journey, heavy food was avoided. Many ships had capsized in this region. Frankly speaking, those 3-4 days of turbulence, crossing the Roaring Forties was more deadly or hazardous than the scariest or deadliest of rides in any theme park.

going to the toilet was the biggest difficulty. We were swaying wildly while walking as if we were in space, like astronauts or cosmonauts. The rolling and pitching was so intense that some members used to fall down from their beds in the night and a few of them had to tie themselves tightly to the bed with belts or ropes. Sometimes, in the middle of the night wardrobes in our cabin would suddenly fling open with a big bang resembling the horror sound effects in suspense thrillers. And all our belongings would be strewn on the floor.

It was difficult to sit down to a meal on the dining table, as many a times the filled plates and glasses would fall on the floor due to impact of rolling and pitching. By the time one would clean up the floor a few utensils would fall in the kitchen. Walking with the food plate was arduous and eating on the dining table impossible. So the tables and chairs were tied tightly inside the kitchen.

Entering the Southern Ocean: As we moved from more turbulent latitude towards the Southern Ocean the sea was gradually becoming calmer and a cool breeze started blowing from the sea raising our curiosity to see something different, especially the ice. A feel of entering into the waters of the Southern Ocean was felt when I saw a distinct zone of blue water merging with slightly milky cold water and pieces of ice floating on the water surface occasionally. The ship started slowing down for taking various oceanographic data and stopped in between for collection of "Antarctic Krill" *Euphausia superba*. These krills are basically shrimp or prawn like crustaceans which are very rich in proteins.

The ship slowly entering the Southern Ocean.

Seeing the first iceberg.

As we moved further down south the captain announced from the deck that an iceberg was on its way. Immediately I was ready with my camera and binoculars to have the first glimpse of an iceberg and capture it in my camera. I was very excited to see such a huge mass of white mountain-like structure. As we moved nearer to the iceberg, I felt like jumping on it or at least touching it with my hand. I had never seen such a huge mass of ice in my life. As we moved further I saw many icebergs of different shapes and sizes on the way. I was so excited that I wanted to go very close to these icebergs. I asked the captain of the ship if he could take us near the iceberg. He told that it was dangerous to be too close to an iceberg as it could cause shipwreck. That time, I did not believe him, but a few years later when I saw the movie *Titanic* I got my answer. An iceberg can weigh from 20 lakh tonnes to 4 crore tonnes.

In the meantime our days were gradually lengthening and night times were reducing to only a few hours making me realise that our ship had already entered "The Antarctic Convergence" i.e. the area where the cold water of Southern Ocean meets with the warmer water. This is one of the most productive ocean zones in the world. The Antarctic Convergence is a shifting zone, and not a precise line which is located between 50° and 60° S. We crossed 50° S on our eighteenth day where atmospheric temperature was 3.5 C. On the twenty-first day (temperature was –3C) as we were passing through 60° S the ocean was frozen to ice and the ship was cutting the "pack ice" which was breaking into large ice flakes. As we moved further slowly, the thickness of ice was increasing and our ship was moving with an amazing sound of "oooh- aaah" giving us frequent jolts. The thickness of sea ice was gradually increasing and the passage was surrounded by many icebergs

Right: Helicopter guiding the ship.

By breaking the thick sea ice, the ship creates an easier path for other ships to follow.

The ship breaking the thin layer of frozen ice into ice flakes.

making navigation difficult and risky. For the next three days the ship was taking a zigzag route to avoid any damage to the ship. The captain told us that the previous year it had been easy to navigate as they encountered the first pack ice at 65° S while this year we encountered it at 60° S. It was an amazing sight to watch the ship cutting the ice and moving at a snail's speed, and I was rushing with my camera from the pilot's cabin on the top floor to the front of the deck to capture as many images as possible.

BROB

December 20, 1987, 7.00 am: Finally, after 25 days of non-stop sea journey covering a distance of 11, 173 kms, we could see the Antarctic continent from a distance!

Transportation is the most hazardous and difficult operation in any Antarctic Expedition. Damage to vehicles was always a possibility and repairs had to be done immediately with available resources.

I was so thrilled that I came back to deck again and again after gaps of a few minutes each to have glimpses of the continent and the penguins. These photos exhibit our welcome party: Emperor penguins and King penguins holding court and welcoming visitors to their native land frolicking and showing us that life in extreme conditions can be fun too.

It was 7:00 am on December 20, 1987, when finally, after twenty-five days of nonstop journey, covering a distance of 11,173 km I could see the Antarctic continent from a distance. Two hours later two pilots from the Indian Air Force and Navy took the first helicopter sortie on a Chetak helicopter to find out the nearest location to *Dakshin Gangotri* so that the ship can be anchored to ice shelf. The captain was moving the ship slowly inside the Polynia to look for anchoring space for the ship. By this time we had crossed 69º S. Finally the ship anchored at 69.3º South Latitude 12.5º East Longitude to the ice shelf at 8:00 pm and the captain announced that we have reached our destination–the Antarctica and the ship cannot move further. After hearing this good news all of us rushed to the deck to have the first closest glimpse of this great continent. A number of penguins were there to welcome us to the continent. It was a beautiful sight to watch these penguins who were flapping their wings and standing in a queue in a highly disciplined manner. After an hour we left the deck and moved to our respective cabins. I was so thrilled that I came back to deck again and again after gaps of a few minutes each to have glimpses of the continent and the penguins.

I did not feel like sleeping although it was midnight. The sun was very bright and the feeling was like as if we were relaxing on a calm Sunday afternoon during Delhi's summer time. My sleeping pattern had already been disturbed since the last few days as it was continuous daylight for twenty-four hours. The next morning when I got up I saw that the penguins were still there. I was very eager to get down from the ship and hug them but there was no way. So I volunteered in the unloading of cargo and supply materials which needed to be transported to the station.

Facing page:
Penguins on the way to welcome us.

Sahoo holding a penguin.

Unloading of items gave me the first opportunity to set foot on Antarctica on December 21, 1987 i.e. after twenty-six days of my journey. I walked very close to the penguins but they were not scared off; instead they became friendly. It was a very exciting moment.

After two days of wait, in the ship, I finally boarded the Chetak helicopter at 8:30 am and left for *Dakshin Gangotri*. My palpitations began to increase as the helicopter was airborne from the ship's deck and flew over the Antarctic Ocean for a brief period and then to the interior of the continent. While flying you have to take all the precautions and wear a water proof "life saving suit". This was mandatory because if by any chance the helicopter crashed into the ocean your body needed to be protected from the freezing water entering inside. (A helicopter had crashed in the Antarctic Ocean on December 29, 1983 during the third Indian expedition) .

While in Antarctica, your whole body needed to be covered. A thick cap on the head, a few layers of woollens on the body topped up by windcheaters. You must wear thick gloves, polar socks, ice boots and sunglasses to protect your eyes. The only exposed parts of your body will be the nostrils and may be the lips. The reflection of ice is extremely high, so if you see the ice with the naked eyes you might lose your eyesight permanently.

Standing on the roof of *Dakshin Gangotri* station.

Facing page:
Life saving suit which is mandatory during the flight (left); arriving at *Dakshin Gangotri* (right).

Although the suit was heavy and uncomfortable one had no choice but to wear it. After flying for few minutes over the snow desert I was able to see a half buried building painted in the Indian tricolour on its roof. There was no trace of any other thing except the white snow as far as the eye could see, right until the horizon. I have no words to describe that view and experience. The weather was very beautiful with a bright blue sky and clear visibility, but the pilots cautioned me that everything was uncertain in Antarctica. The weather might change any time, and if that happened we might not be able to land at the station and might have to go back to the ship. (Weather in Antarctica is highly unpredictable and in no time the clear sky fades and a blizzard can come with a velocity of more than 100 km/hr making life difficult without clear visibility/with near zero visibility). Anyhow, that day God had been very kind to us. There was no blizzard and the helicopter landed at *Dakshin Gangotri*. I came out of the helicopter and kneeled down on the ice facing *Dakshin Gangotri* and kissed the ground with all humility. It was an amazing sight to see a wonderful building half-buried in the snow and there was no other structure nearby. I went inside the station and was heartily welcomed by the members of the wintering team who had been staying there for nearly a year. I spent a few hours interacting with the wintering team members to learn from their experience, so that I could be prepared and get familiar with my work in the forthcoming days. Finally, after planning my experimental setup, I retuned back to the ship the same day in an Air Force helicopter.

On December 24, 1987, on the eve of Christmas the captain of the ship along with his entire Swedish crew members invited all of us for a Christmas celebration. There was a gala party with music playing full blast and sixty-four varieties of Swedish dishes being served. The captain presented us with Santa Claus's gift. (To be very honest I had never seen that many varieties of food being served before). All of us had a good time. That evening we saw a huge Russian Aeroflot Cargo plane hovering around the ship. It was very exciting to see an aeroplane after many days. Later we found out that the plane landed on ice near the Russian station "NOVO" carrying the station supplies and team members.

During our journey, we were told that while in Antarctica, one had to be always alert and active. One should get ready in 5-10 minutes to fly or do any work assigned to them. After Christmas Day the weather remained pretty bad for two and a half days and there were no activities due to heavy blizzard. Around 4:00 pm on December 28, 1987 the sky became clearer and the leader told me to get ready in ten minutes to fly to *Dakshin Gangotri*. Immediately, I put on layers of clothes, picked up my backpack and boarded the helicopter. By 4:45 pm we landed at *Dakshin Gangotri*. For next few hours we started shifting heavy equipment of Indian Meteorology Department near the station. Then I, along with another colleague, moved to a fully underground (buried) container of Defence Research Development Organisation (DRDO) for our temporary accommodation which was located hundred metres away from the station. Around 9:00 pm (broad daylight) a heavy blizzard started, bringing life to a standstill. For the next few days the blizzard was continuing in full

After blizzard the ice becomes very sharp like razor due to impact of bright sunlight.

force with a visibility of hardly five-ten metres. There were no toilet facilities inside our underground container. With a lot of difficulty one had to come out of this container and walk in the blizzard bending the body to an almost 30° angle. The impact of this snowstorm was so high that many times you had to walk backwards to avoid accumulation of snow particles on your moustache and beard. Besides, exposing your backside while you relieve yourself was one of the most challenging and difficult tasks because of the extreme low temperatures and high wind speed. This was the most terrible experience for most of us.

Anyhow on December 31, 1987 the station commander of *Dakshin Gangotri* invited us for the New Year celebrations. With utmost difficulty we walked in the blizzard from our cabin to the Indian station and started the New Year celebrations exactly at midnight. Welcoming the New Year 1988, in Antarctica, was a unique opportunity and an unforgettable experience.

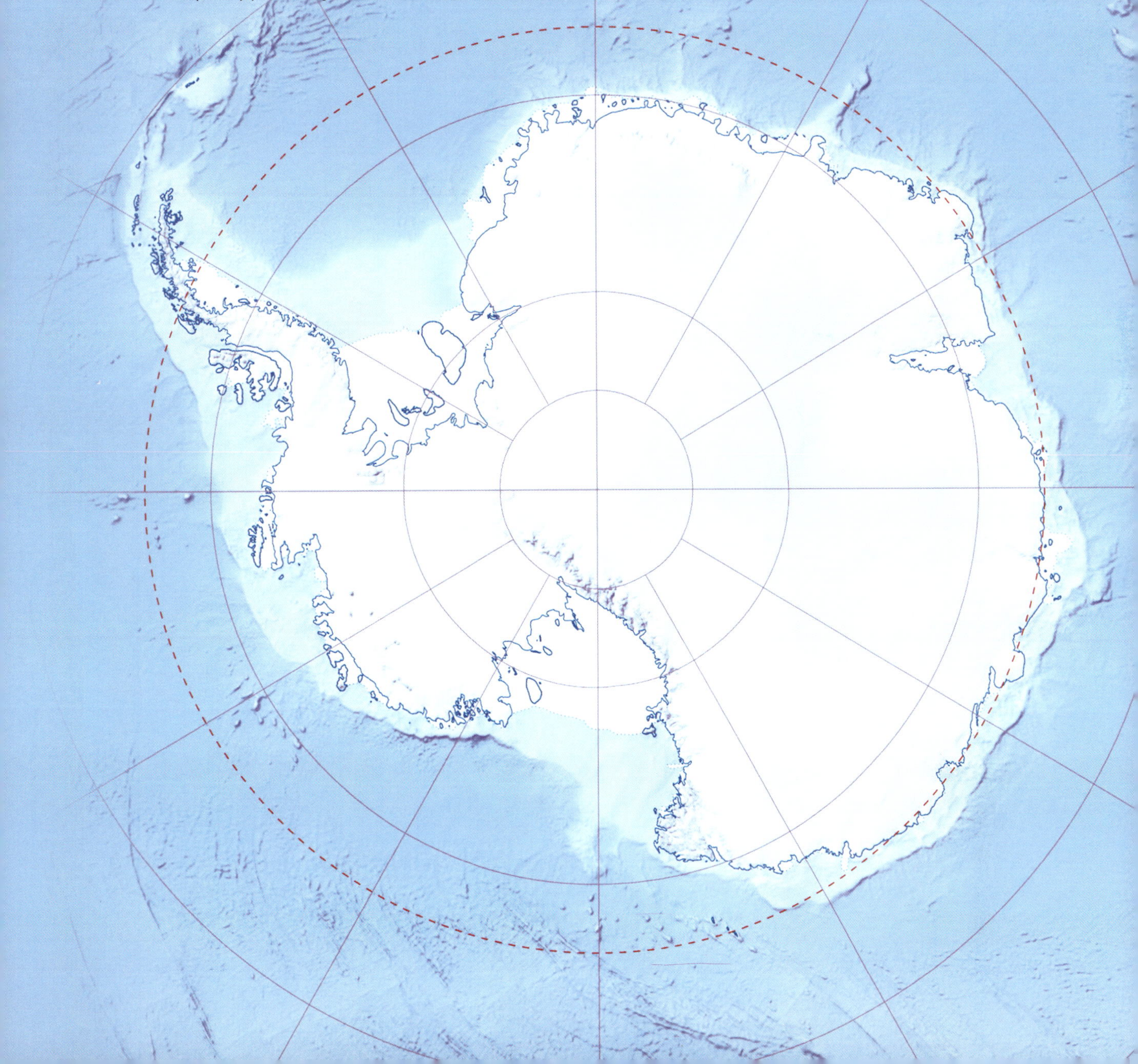

Map of Antarctica.

The interior of the station was well decorated like in any other New Year celebrations. While having the party, somebody said, "Can you get me some ice cubes for my drink?" Pat came the reply, "Why do you need ice cubes, there is ice-ice-and-ice everywhere around." Someone else went out and filled up a bucket with fresh snow as a substitute for the ice cubes. This was all really fun.

Walking in blizzards was fun as well as dangerous. I spent the New Year of January 1988 walking intermittently around *Dakshin Gangotri*. It was fun to watch the deposition of snow on the roof of the station as well as on the snow vehicles parked outside. The real challenge came when one needed to convert the snow into drinking water as well as during the removal of snow from the station's roof and entrance. Although Antarctica has 70% of the world's fresh water deposits, getting water in Antarctica was a luxury and an arduous task.

Hectic activity at *Dakshin Gangotri* during setting up of Maitri.

In our station, there was a snow melting plant which provided water for drinking, cooking, washing and the occasional bath by turns. Every member, irrespective of bad or good weather had to go out of the station and manually lift snow and put it into the snow melting plant. This used to be a very tough job but there was no choice if one wished to survive.

Similarly, every blizzard used to cover the station with heavy snow. We had to remove the snow from the surface of the station, otherwise it would get buried. (I must confess here that this process was more tough during Antarctic winter. In the polar nights I performed this task several times in the middle of the night). Anyhow, after six days of staying in the field and being cut off in the blizzard I flew back to the ship.

109

The biggest enemy during our stay in Antarctica was not any animal but psychological pressure, loneliness and blizzards. The vast expanse of ice was so huge, that if you lost direction during your exploration, there would be nobody to guide you back. If you were lost, you remained lost forever.

Celebrating Republic Day

On January 26, 1988, all members flew to *Dakshin Gangotri* to take part in the Republic Day celebrations. Members of the Russian and German stations as well as our Swedish crew members were invited to the Republic Day Parade. At 11:00 am the Indian tricolour was unfurled followed by the singing of our National Anthem. My heart was overwhelmed while taking part in this historical moment, an unforgettable event in my life. Flags of participating institutes were hoisted and I had the privilege of hoisting the flag of Delhi University in Antarctica.

During my stay at *Dakshin Gangotri* I braved many days of blizzards and extremely harsh weather conditions. Walking on the ice was not only difficult but also very risky. There were loose ice patches and in many places there were hidden crevices. If one was not careful during walking or driving

Celebrating Republic Day at Antarctica on 26 January 1988.

a snow vehicle, there was every possibility that one would fall inside these crevices. If that happened nobody could have been saved. So even if one had to walk a distance of fifty or hundred metres on the ice, it was necessary to walk with the help of an ice axe and take a zigzag turn to avoid crevices. To cover a distance of hundred metres, many a times it used to take one full hour or more.

Establishing *Maitri* India's second permanent station in Antarctica: One of the major objectives of our expedition was to construct India's second permanent station *Maitri* at Schirmacher Oasis, an ice free region located nearly hundred kilometres further south of *Dakshin Gangotri*. This is a beautiful picturesque location with mountain ranges and a number of lakes. After staying in the ship for nearly five days due to bad weather conditions, finally I flew down to Schirmacher Oasis in my third attempt.

Inset: Sahoo standing with the Delhi University flag.

Schirmacher Oasis, an ice free region located nearly hundred kilometres further south of *Dakshin Gangotri* is a beautiful picturesque location with mountain ranges and a number of lakes. This was chosen as the site for India's second permanent station *Maitri*.

During our stay at *Maitri* our first job was to construct two helipads so that the helicopters could land and take off safely. We cleared the ground of unwanted rocks and boulders and then levelled it. Utmost care was taken not to disturb the precious plant and animal lives in the surrounding area. Our building construction in-charge was a very tough, well built and disciplined army man. Thus our daily routine was very tough. Besides carrying out our scientific experiments and exploration, it was mandatory for all the scientists to be part of logistics team and work hard with others from 9.00 am to 5.00 pm irrespective of the weather. (No work then no food and no sleep was the martial order regardless of your scientific work. We were able to complete the building in a very short period only because the construction in-charge was a rigorous hard taskmaster. Shifting of oil drums, loading and unloading of heavy construction material, digging the ground, breaking stones, preparing concrete mixtures, lifting the pillars on the head and shoulders and erecting them at–10° C with high wind velocities was a tough job. Furthermore, another uphill task was to lift water from the frozen lake by breaking the upper layer of ice and carrying it across a distance of five hundred metres on a tough steep terrain to the construction site. Although many of us were hesitant and reluctant to carry out the construction work and grumbled, today I cherish those moments and feel proud to have been a part of history in laying down the foundation of India's second permanent station.

Facing page:
Site for *Maitri* (above); laying foundation for *Maitri* during heavy snow (middle); taking oil to heat the cottage (bottom).

In the night, around nine, when some of us were busy fixing the roof of the new station, suddenly the blizzard started picking up speed. It was very difficult to stand on the rooftop of the building fixing the roof. While we were climbing down, a huge piece of the wooden roof was blown away and fell on one of the army jawans standing down below. The jawan fell down on a heavy

Temporary hut laid down during the previous expedition at the construction site facing the Indira Priyadarshani Lake named after the then Prime Minister.

SITE
FOR
INDIAN
STATION

Above: Discussing about the construction (left); bonfire (middle); after completing one block the foundation for the second block is being laid (right).

Laying the helipad for the construction of *Maitri*.

NAVY
464

boulder and started crying out with pain. He was immediately brought to the base camp for necessary treatment, but had to be content with only hot water treatment and some pain relieving ointment as first aid. In the meantime the wind speed had picked up to 80 knots per hour with temperatures falling down to -18°C. The communication system from the base camp to the ship had broken down and there was no way to contact the doctor for emergency medical help. Added to this medical emergency, the wooden huts in our base camp were rattling accompanied by the cracking and whistling sound of the blizzard. These were extremely scary nights in the field camp.

Finding some leisure time from our difficult schedule, one afternoon, I along with an army logistics personnel visited the nearby Russian station "Novolazarevskaya" (Short name "Novo") which was located five-six km away from *Maitri* and could be seen from the hilltop. We were warmly greeted by the members of the Russian station with their traditional offering of bread and vodka. Two members from the nearby German station also joined in the discussions. It was a great feeling to meet people in this no man's continent. After having a good time we started to trek back to our base camp at around eight. As this was mid-February the sky was getting slightly dark compared to the bright nights in the month of January. Suddenly we saw that the weather was changing rapidly with high wind speed and reduced visibility. We guessed that a blizzard will hit the area very soon. So we started running, but it was difficult. In the meantime we lost our way to the base camp and landed somewhere near a birds' nesting ground. We found a large number of skua birds with a few newly hatched ones. As we tried to pass the nesting

Novolazarevskaya station Helicopter Photo.

A skua in an attacking mode.

grounds, a number of birds started hovering over our heads with high pitched sounds and began readying to attack us with their beaks. Probably the birds thought that we had come there to take away their newly-born. So immediately we started running away from the nest sensing danger and the birds calmed down. It was quite a harrowing experience. We continued our journey towards the base camp but there was no trace of any sign even after two hours of trekking. It was getting darker and the wind speed was slowly increasing with the furious whistling sound. Our heart was beating fast as if racing to an unknown destination with deep fear in our minds. I thought we were not going to make it and were destined to die there. Although both of us were extremely tired and our body had given up, we started praying to God to show us the direction. In the meantime the other team members at the base camp had started getting worried as they had no contact with us. Finally around midnight we were able to see our base camp from a distance and started running. Everybody was relieved to see us back. The blizzard picked up speed and we remained confined to our station for a week. Food stocks were getting over day by day; we were waiting for fresh supplies from the ship but in vain. As the expedition was coming to an end and we were to leave Antarctica by February 24, 1988, we were advised to finish the work as soon as possible and return back to the ship. Finally, the sky slowly cleared after a week of blizzard. We had been waiting eagerly to leave *Maitri* station by the fist available chopper. I left *Maitri* station on the afternoon of February 22, 1988 and landed at *Dakshin Gangotri* to pick up my samples and other belongings. Due to the previous week's blizzard the whole station was buried under a thick blanket of snow and only the chimney was visible. With the help of a ladder we could climb up the chimney and made an entry inside the station climbing down from the chimney. After thirty minutes, when we came out, the helicopter had already left and did not come back due to bad weather. So we returned to the underground bunker for our night stay. As the weather was getting better and visibility had improved, at night we could see the lights of the ship from *Dakshin Gangotri*. It was such an enchanting scene to watch the twinkling lights of the ship from a distance of nearly fifty km. This view was possible as the atmosphere was very clear and the ice surface was completely flat. The next day morning I left the station and landed in the ship.

Good news! I had been selected as the First Indian Student to take part in the Seventh Indian Antarctic Expedition! This was a very happy moment in my life, no, the happiest one, a once-in-a-lifetime experience! As great as my co-author Dr S.Z. Qasim's on leading the First Indian Expedition and being the First Indian to set foot on the continent!

The icebergs, the penguins their discipline and obvious sense of joy in living, the unpredictable weather conditions, the picturesquely clear bright blue sky, the transportation of foodstuff and materials, the participating in construction activity, tent life, were all images we wanted to capture and store for posterity.

The snapshots in this page will bear out our excitement and make our readers want to visit Antarctica.

Last glimpse of the continent during departure.

Finally I could take a nice hot water shower in the ship after nearly two weeks and have some good food. The next day, when we were about to depart from Antarctica, we found that one of the logistics personnel was missing. After a frantic search we located him at *Dakshin Gangotri* , and a helicopter was sent to bring him back to the ship. As the ship was about to leave, the weather started taking a turn for the worst. So we had to wait for a few more days. The patience of the members was running out. Each one of us was feeling homesick and wanted to return to civilisation. Eventually, the weather god started smiling on us. At last, on February 28, 1988 at 3:26 pm the ship departed for India and a smile lit up everybody's face. As the ship moved, it took a different route. We

Today, I consider myself fortunate and am proud to have taken part in laying down the foundation for India's permanent research station called *Maitri*. Constructing a building in harsh and extreme weather conditions was a difficult job. Keeping very high spirits wanting to be part of history, I joined hands with a few other members in constructing the building. After the coconut was broken we dug the rocky ground, lifted the iron planks on our heads and shoulders to the construction site. We broke the stones, prepared the concrete mixtures, and erected the iron pillars on the ground. We also undertook and carried out the most difficult job of carrying water buckets from the lake to the construction site across a distance of 300-500 metres. Today after more than twenty-five years, the building remains strong. Not only that, it is the epicentre of Indian Scientific activities in Antarctica

were told that we would pass through "Kergulen Islands" (a French territory) and will not go through Mauritius which was the normal route. As we continued our journey the seawater started freezing into ice.

We crossed 60° S on March 4, 1988, 50° S on March 8, 1988. We crossed the roaring 40° S on March 10 and the equator on March 22, 1988. So, after spending 121 days on the high seas and the frozen continent , we returned to India on March 1988 and were greeted at Goa with the usual ceremony. Those 121 days of expedition had left an everlasting impression on my mind and were a wonderful, unforgettable experience in my life, one which cannot be explained, but has to be felt. I lived every moment of it and even after two decades or so I would love to visit this wonderful creation of nature once again.

Maitri now is an international station, manned round the year, capable of accommodating about twenty-five persons comfortably, well furnished with living space, laboratories and modern amenities, which offer excellent infrastructure for carrying out front-ranking research in several areas of science. Communication between India and *Maitri* is well-established through satellite, HF linkages, and telephone, fax and telex systems.

Antarctica gave me everything unique This expedition had given me many things in my life. In those years, the 1980s many things were difficult such as getting a flight ticket (only Indian Airlines and Air India used to operate those days), getting a passport, visiting foreign countries, etc. I must confess that the first time I boarded an aeroplane in my life was in September 1987, when

Left: Rajiv Gandhi talking to expedition members.

Sahoo with Rajiv Gandhi before proceeding to Antarctica.

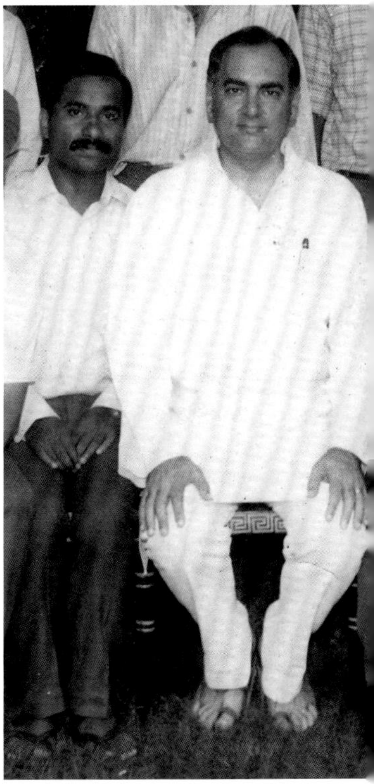

Above: Sahoo receiving a memento from K.R. Narayanan the then Science & Technology Minister, India.

Facing page: Sahoo with R. Venkataraman, the then President (above); Sahoo and Mrs Sahoo with Rajiv Gandhi after coming back from Antarctica (back).

I flew from Delhi to Srinagar for my training at High Altitude Warfare School. For the first time I flew in a helicopter. When I applied for my passport the regional passport officer was very kind enough to help me, when he heard that I will be going to Antarctica. I undertook my first overseas trip to a place where you do not need a visa. For the first time I travelled in a ship. I took some dollars but could not buy anything in Antarctica as there were no shops. Out of so many firsts in my life, the most important First was the first time I

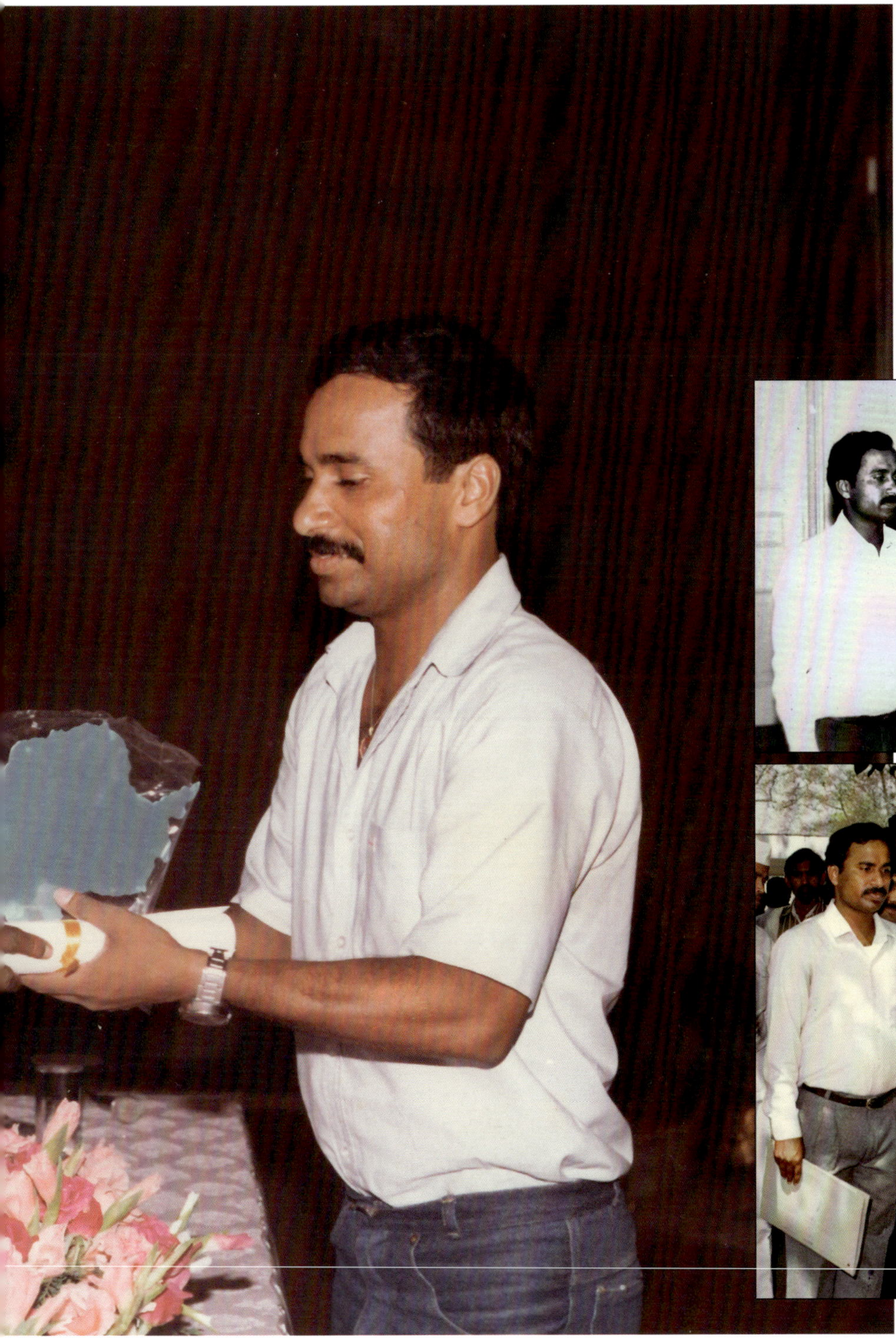

was imbued and being overswept with the emotional and patriotic feeling that I was doing something special for India. The feeling of being an Indian, that special feeling for the country was truly unique, one I could feel at the onset of the expedition. Another high was the feeling that I celebrated the New Year and India's Republic Day in Antarctica.

Benefits from Antarctic Research

Benefits from Antarctic Research

Benefits from Antarctic Research to the country are many. First and foremost, the Antarctic programme had made it possible for India to build a team of persons capable of working in extreme cold and difficult conditions. By their untiring efforts, first-hand knowledge about Antarctica has been obtained. During the expeditions, the scientists had been able to test numerous indigenous equipment and other material in Antarctic conditions which would not have been possible otherwise.

A major breakthrough was made through the use of containerised accommodations in Antarctica developed by the Defence Research and Development Organisation (DRDO). As a result of the experience gained, container-type living accommodation has improved considerably over the years. The experience gained in Antarctica has been successfully utilised in strategic areas for India's defence needs. Rich experience has also been gained in the methods of erection and construction of buildings in extremely hostile weather conditions. The design and development of special clothing, footwear, tents and other accessories developed for living in Antarctica are now being used in the frontier areas by Indian defence forces.

An important benefit had been in the establishment of long-distance communication links using indigenous equipment, the construction of helipads, operation and maintenance of vehicles and indigenous methods

Facing page: *Bharti*- India's third permanent station.

Dakshin Gangotri India
Maitri India

for the development of various items of processed and cooked food. It has also been possible to acclimatise Indian men to operate vehicles in difficult and unknown terrain. Similarly, the treatment and disposal of waste material, according to international standards, had been an entirely new experience. In the field of meteorology, benefits have been significant. It has been possible to make timely forecasts and issue warnings both for navigation and scientific investigations in Antarctica. Another significant benefit is the understanding obtained about the possible relationship between the seasonal variation of surface temperature over Antarctica and the occurrence of rainfall in tropical regions. Glaciological studies undertaken during these expeditions have provided a basis for studying the relationship between Himalayan and Antarctic glaciology. No other country in the world has so far undertaken such a study. As a result of the Antarctic expeditions, India has been able to build up personnel who are capable of navigation and position fixing in very difficult and unknown terrains.

India's continued operation in Antarctica has facilitated the cooperation of scientists in data and information exchange related to Antarctica with other countries which have an interest in Antarctica. India's Antarctica station provides logistic support to various scientific teams in the form of timely forecasts and warning. A weather forecasting system has already been established by India in Antarctica. Information and data are being collected in respect of general weather conditions and these are being shared with other weather stations in Antarctica. Baseline data for the prediction and forecasting of the Indian monsoons is now available with the Indian Meteorological Department (IMD) in Delhi and the National Institute of Oceanography (NIO), Goa.

Establishment of India's capabilities to measure extremely low levels of radioactive radon in the Antarctic area are also underway. Studies on human adaptation and adjustment in the extreme Antarctic area are also going on. These studies on human adaptation and adjustment in the extreme Antarctic environment have proved to be most valuable.

Facing page:
Co-ordinates of *Dakshin Gangotri*, *Maitri* and *Bharti*, India's permanent stations at Antarctica. The distance between *Maitri* (second station located at India Bay) and *Bharti* (third station located at Larsemann Hills) is 3098 km.

Dakshin Gangotri (India's first station) has gone underground. Should the ice melt, thanks to global warming *DG* may re-emerge or float in the ocean!

Antarctic Science of National significance

Antarctic Science of National significance

India cannot afford to ignore the results obtained from its Antarctic mission in certain fields. In particular it is possible to pinpoint those studies which demonstrate the role of Antarctica in global processes and which have a direct bearing on the welfare of human beings. In the post cold-war scenario, with the closure of the former East German station *Georg Foster* and the reduction in scientific activities of the Russian station *Novolazarezvyka*, the Indian station assumes pre-eminence as the only active base in the Queen Maud Land. Hence, the scientific data collected by Indian scientists assume greater importance. The significance of the researches to India can be described as follows:

1. Monsoons have a direct bearing on the economy of India. The monsoon is controlled by numerous parameters generated over the entire global systems. Climatological data collected by the Indian scientists over a period of years will help to develop a more realistic model on monsoon prediction and to understand the dynamics of the weather system over the Indian Ocean. Similarly, the study of ice cores will give a picture of past climatic records which can provide vital inputs to its prediction model.
2. Antarctica provides a singular platform to monitor ozone depletion. The understanding of ozone-hole phenomenon, despite being a part of the global effort, holds significant interest for Indian scientists. The hole,

Facing page: Penguins forming a Welcoming Party.

which was once considered static, has proved to be dynamic. This will probably have some bearing on the weather patterns of the Indian Ocean.

3. In the geological past (between 700-2500 million years ago), India and other southern continents belonged to a single landmass called Gondwanaland, of which Antarctica was the centrepiece. During this period, an intense magnetic activity took place resulting in the richest mineral deposits that we find in the world today. The endeavour of Indian geologists is to reconstruct the best fit of Antarctica with the peninsular margins of India. This will not only lead to a better understanding of the tectonic history of the earth but will simultaneously help to elucidate the spatial distribution of geo-resources in the cratonic shields of India and Antarctica.
4. The Southern Ocean is a major sink of carbon-di-oxide from the atmosphere. This in turn contributes to the high productivity of this water mass that has a direct effect on the nascent fisheries. Consequently, the stocks of krill, fish and squid in the Antarctic waters offer economic resources in excess of all the oceans put together for sustainable exploitation. The Ocean sector of the Antarctic waters can be a future source of marine living resources which has so far remained under-exploited.
5. Antarctic extremophiles have attracted the attention of Indian microbiologists to exploit their potentials in biotechnology. Attempts are being made to identify low temperature genes available in the Antarctic micro-organisms that can be cloned into mesophilic bacteria. This will help in the disposal (by biodegradation) of human and other organic wastes under extremely low temperature conditions like in the colder regions of the Himalayas.
6. Adaptation techniques of human beings to the harshness of Antarctica and the ways and means to enhance their physical productivity under stress and duress will give a vital insight in improving the mental and physical health of our defence personnel living in similar conditions in the frontier regions.
7. An understanding of the plasma processes occurring in the Earth's upper atmosphere will have practical importance in relation to long distance radio communication and satellite operation. The high latitudes, particularly Antarctica, have provided a region uniquely favourable to study the best possible windows in communication especially for high frequency and very low frequency propagations.
8. Antarctica offers an unparalleled laboratory to develop expertise in low temperature structural engineering. Technologies developed in the field of life-support systems and station infrastructure put up at *Dakshin Gangotri* and *Maitri* have been successfully applied for designing and setting up of advanced bases in the Himalayas.
9. The Indian Army, Navy and Air Force, by virtue of their participation in the expeditions, have gained professional expertise of the highest order in snow-ice navigation under adverse climatic conditions. Together with this, the capacity to establish an effective communication link between the Antarctic station and the mainland, through high frequency, has already been demonstrated.

Geopolitical element

While on the scientific side, India now has a very strong base, it is equally interesting to note that there is a strong geopolitical element imbibed in the Antarctic programme. It may be recalled that although the origin of Antarctic Treaty of 1959 dates back to the International Geophysical Year, a closer examination reveals that it began with the first International Polar Year. It was during these years that the latitude and longitude of the Antarctic region were determined. The preparatory work for the Antarctic Treaty was carried out during the International Geophysical Year through a Special Committee which, at that time, was known as Standing Committee on Antarctic Research. This Committee was renamed as the Scientific Committee on Antarctic research (SCAR) in the year 1961, when the Antarctic Treaty came into force. Thus SCAR and Antarctic Treaty combined or in other words science and politics were intermeshed under a broad framework. At present, all the activities and affairs connected with the Antarctic continent including its environment and conduct of scientific research carried out there are managed by the Antarctic Treaty System (ATS).

The major objectives of the Treaty are:

1. use of Antarctica for peaceful purposes only;
2. facilitation of scientific research in Antarctica;
3. facilitation of international scientific cooperation in Antarctica;
4. facilitation of exercising the right of inspection as provided in Article 7;
5. forbidding the question relating to the exercise of jurisdiction in Antarctica;
6. preservation and conservation of living resources of Antarctica.

The sea ice is perhaps the most enthralling of all, its beauty changing with the seasons. It forms in the autumn, expands widely in winter and begins to break up in spring. In summer it drifts as pack-ice until much of it melts away as it moves into somewhat warmer waters to the north.

Recognition of India's capabilities

Recognition of India's capabilities

To gain access and be a party to the operative mechanism, a new contracting party could be given a consultative status if it demonstrates substantial scientific interest in Antarctica or engages in expeditions to Antarctica. India's entry to the Antarctic Treaty mechanism and immediate grant of consultative status in 1983 is a matter of great satisfaction and an attestation by the international community of India's scientific capabilities and accomplishments. Countries which are the elected members of the Treaty are called the Antarctic Treaty Consultative Status Parties (ATCSP). At present, there are twenty-six such ATCSP member states including India. In addition, there are sixteen other states which have acceded to the Antarctic Treaty but have not yet attained the consultative status. India also opened a token post office in Antarctica to show its presence.

A postage stamp is being released to mark the first expedition.

Facing page: India's first post office was opened during the seventh expedition 1987-88.

Citizens or Residents of Antarctica

Citizens or Residents of Antarctica

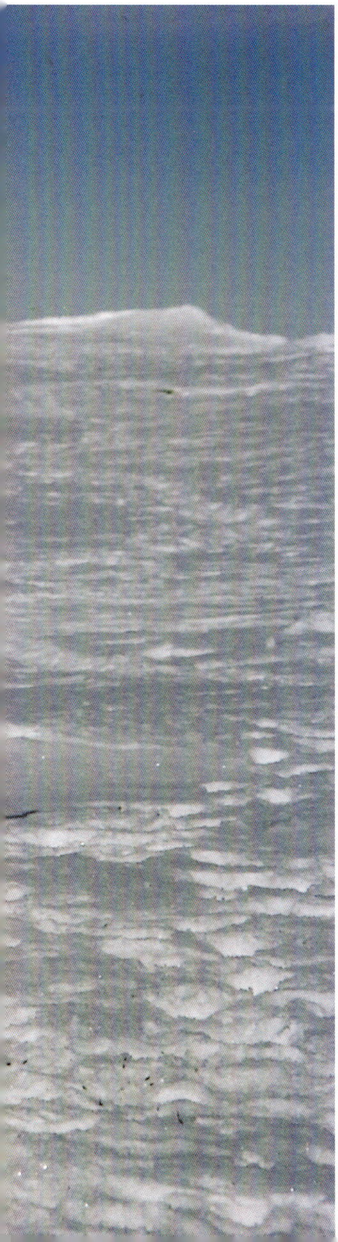

Antarctica has no permanent residents, but a number of governments maintained permanent manned research stations throughout the continent. The number of people conducting and supporting scientific research and other work on the continent and its nearby islands varies from about 1,000 in winter to about 5,000 in summer.

The first semi-permanent inhabitants of regions near Antarctica (areas situated south of the Antarctic Convergence) were British and American sealers who used to spend a year or more on South Georgia, from 1786 onwards.

The first child born in the southern polar region was a Norwegian girl Solveig Gunbjørg Jacobsen, born in Grytviken on 8 October 1913, and her birth was registered by the resident British Magistrate of South Georgia.

Emilio Marcos Palma was the first person born south of the 60th parallel south (the continental limit according to the Antarctic Treaty), as well as the first one born on the Antarctic mainland, in 1978 at Base Esperanza, on the tip of the Antarctic Peninsula; his parents were sent there along with seven other families by the Argentine government to determine if family life was suitable on the continent. In 1984, Juan Pablo Camacho was born at the Frei Montalva Station, becoming the first Chilean born in Antarctica. Several bases are now home to

Facing page:
Antarctica's true and faithful citizens.

families with children attending schools at the station. As of 2009, eleven children were born in Antarctica (south of the 60th parallel south): eight at the Argentinean Esperanza Base and three at the Chilean Frei Montalva Station.

Antarctica has no government, although various countries claim sovereignty in certain regions. While a few of these countries have mutually recognised each other's claims, the validity of these claims is generally not recognised universally. New claims on Antarctica have been suspended since 1959 and the continent is considered politically neutral. Its status is regulated by the 1959 Antarctic Treaty and other related agreements, collectively called the Antarctic Treaty System. Antarctica is defined as all land and ice shelves south of 60° S for the purposes of the Treaty System. The treaty was signed by twelve countries including the Soviet Union (and later Russia), the United Kingdom, Argentina, Chile, Australia, and the United States. It set aside Antarctica as a scientific preserve, established freedom of scientific investigation and environmental protection, and banned military activity on the continent. This was the first arms control agreement established during the Cold War. Thus, Antarctica has no official flag as it is not a nation nor is it ruled by a single government or body. The Antarctic Treaty Organisation adopted a flag in 2002 and this is used as an unofficial symbol of the continent.

In 1983, the Antarctic Treaty Parties began negotiations on a convention to regulate mining in Antarctica. A coalition of international organisations launched a public pressure campaign to prevent any minerals development in the region, led largely by Greenpeace International which established its own scientific station–World Park Base–in the Ross Sea region and conducted annual expeditions to document the environmental effects of humans on the continent. In 1988, the Convention on the Regulation of Antarctic Mineral Resources (CRAMRA) was adopted. The following year, however, Australia and France announced that they would not ratify the convention, rendering it dead for all intents and purposes. They proposed instead that a comprehensive regime to protect the Antarctic environment be negotiated in its place. The Protocol on Environmental Protection to the Antarctic Treaty (the 'Madrid Protocol') was negotiated as other countries followed suit and on January 14, 1998 it came into force. The Madrid Protocol bans all mining in Antarctica, designating the continent as a 'natural reserve devoted to peace and science'. The Antarctic Treaty prohibits any military activity in Antarctica, including the establishment of military bases and fortifications, military manoeuvres, and weapons testing. Military personnel or equipment are permitted only for scientific research or other peaceful purposes.

Penguins and more penguins! We are the lords of all we survey!

Conclusion

Since the First Indian Expedition landed in Antarctica, thirty years have passed. During this period thirty-one expeditions have landed in Antarctica at the rate of roughly one expedition each year. During the year 2000, launching of the Indian expedition from Goa was discontinued. Thus from the twentieth expedition onwards, the members of the Indian Expedition fly to Cape town in South Africa and from there the team travel by the ship which takes around 10 days. India's third permanent station *Bharti* is being now established in the year 2011-12 at the Larseman Hill. The frozen continent of Antarctica at which we looked thirty years ago with mystery, awe and fear is no longer mysterious or intimidating to Indians. The Indian scientists live and work most comfortably in this continent throughout the year and carry out their work during the winter months when at times the temperature drops down to -50˚C. Antarctica is a beautiful part of our Earth–most promising and rewarding–and therefore, must be studied by us and by our future generations. It holds a great future for mankind. We have yet to find a person who has been to Antarctica once and would not like to go again. Such is the charm of that unpolluted and uncontaminated continent and such is the enchanting landscape of the former *Dakshin Gangotri* and *Maitri* and the newest one *Bharti*–calm, serene, pure and most inspiring.

Today, going to Antarctica has become easier. One can go to Antarctica (mostly to sub-Antarctic regions not deep

Facing page: A view of the ship through an iceberg.

into mainland Antarctica) as a tourist for 10-15 days in Polar Cruises either from Argentina, Chile, Australia or New Zealand by paying between US $ 7,000-12,000. You can fly to Argentina and then catch a tourist ship from Ushuaia or from Punta Arenas in Chile or from Christchurch in New Zealand or Tasmania in Australia. Earlier there were tourist flights operating from New Zealand to McMurdo Sound in Antarctica but after an aeroplane crash the service was discontinued. Today, there are a few reputed travel agencies/ companies affiliated to and recognised by International Association of Antarctic Tour Operators (IAATO) which provide responsible travel and tourism to Antarctica. At the end we suggest and appeal to one and all to preserve and protect this greatest wilderness in the world so that the future generations can inherit this nature's best gift to mankind.

Aurora Australis-A full moon and 25-second exposure allowed sufficient light for this photo to be taken at Amundsen-Scott South Pole Station during the long Antarctic night. The green light in the background is the Aurora Australis
Courtesy: Wikipedia.

Bibliography

© Mike Usher

Mosley Geoff (1986). *Antarctica, Our last great wilderness*, Australian Conservation Foundation, Australia.

Readers' Digest (1985). *Antarctica Great Stories from the Frozen continent*, Sydney, Australia.

Qasim S.Z. (1998). *Glimpses of the Indian Ocean*, Universities Press (India) Limited, Hyderabad, India.

Sahoo Dinabandhu and Pandey P.C. (2002). *Advances in Marine and Antarctic Science*. APH Publishing Corporation, New Delhi, India.

Sharma S.S. (1997). *Citadel of Ice, Dakshin Gangotri: The first Indian Wintering in Antarctica*: New Age International, Delhi, India.

www.coolantarctica.com
wikipedia.org/wiki/Antarctica
NASA, National Geography Channel, (map of Antarctica)

Hussain, M.F. 2011. *Syed Zahoor Qasim. The Man, Mission and Milestones* ... by Meher Fatima Hussain, Manak Publications Pvt. Ltd. p. 110.

What a crowd! Is there some meeting or conference going on! Probably the move is to oust humans who are crowding and spoiling their rightful territory!

A view of Mc Murdo Sound station.

Antarctica now! With the crowding in that is happening in the name of research and setting up of research stations by countries from all over the world, Antarctica is in grave danger of losing its reputation of vast wilderness, wide open spaces and thin population where if you lose your way, you may be lost forever!

We fervently appeal to one and all to preserve and protect this greatest wilderness in the world so that the future generations can inherit this nature's unique gift to mankind.

Acknowledgements

We have been thinking of writing this book for quite some time, however it could only take shape last year. While preparing this book we received help and guidance from several people. We are extremely grateful to Professor Paul A. Broady, University of Canterbury, Christchurch, New Zealand for critically going through the manuscript, providing valuable suggestions and sharing some of the beautiful photographs. Our grateful thanks to Mr Mike Usher, Dr Prabhu Matondkar and Major General S.S. Sharma all of whom had been to Antarctica in various expeditions and who have been kind enough to provide some important pictures for the book.

We are grateful to Dr Shailesh Nayak, Secretary, Dr M. Sudhakar, Advisor and Dr N. Khare, Director, Ministry of Earth Sciences, Government of India, New Delhi for providing financial assistance and valuable suggestions. This book would not have been possible without unconditional help and support of our family members Mrs Birjees Qasim, Ms Seeme Qasim, Dr(Mrs) Meenakshi Munshi Sahoo, Mr Chaitanya Sahoo and Ms Anchal Sahoo who gave us the time and space, rightfully theirs, to devote our energies to this book.

We thank Dr Kishore Kumar and Mr Sanjeev Goel from Centre for Ocean and Environmental Studies and Mr Savindra Kumar, Mr M. Arunjit Singh, Ms Salam Sonia Devi and Ms Priyanka Verma research scholars from Department of Botany, University of Delhi, Delhi for their tireless efforts and timely help.

Last but not least; we thank Mr Bikash D. Niyogi, MD, Ms Nabanita Das, designer and Ms Gita Rajan, editor, Niyogi Books for translating our dream into a reality.

A panoramic view of daytime in Antarctica. Let mankind not destroy the beauty of this unique vast White Desert!

Index